HANDBOOK OF RESEARCH ON FOOD SCIENCE AND TECHNOLOGY

Volume 1

Food Technology and Chemistry

HANDBOOK OF RESEARCH ON FOOD SCIENCE AND TECHNOLOGY

Volume 1

Food Technology and Chemistry

Edited by

Mónica Lizeth Chávez-González, PhD
José Juan Buenrostro-Figueroa, PhD
Cristóbal N. Aguilar, PhD

Apple Academic Press Inc. | Apple Academic Press Inc.
3333 Mistwell Crescent | 9 Spinnaker Way
Oakville, ON L6L 0A2 Canada | Waretown, NJ 08758 USA

© 2019 by Apple Academic Press, Inc.
Exclusive worldwide distribution by CRC Press, a member of Taylor & Francis Group
No claim to original U.S. Government works

Handbook of Research on Food Science and Technology, Volume 1:
Food Technology and Chemistry
International Standard Book Number-13: 978-1-77188-718-2 (Hardcover)
International Standard Book Number-13: 978-0-42948-785-9 (eBook)
Handbook of Research on Food Science and Technology, 3-volume set
International Standard Book Number-13: 978-1-77188-721-2 (Hardcover)
International Standard Book Number-13: 978-0-42948-781-1 (eBook)

All rights reserved. No part of this work may be reprinted or reproduced or utilized in any form or by any electric, mechanical or other means, now known or hereafter invented, including photocopying and recording, or in any information storage or retrieval system, without permission in writing from the publisher or its distributor, except in the case of brief excerpts or quotations for use in reviews or critical articles.

This book contains information obtained from authentic and highly regarded sources. Reprinted material is quoted with permission and sources are indicated. Copyright for individual articles remains with the authors as indicated. A wide variety of references are listed. Reasonable efforts have been made to publish reliable data and information, but the authors, editors, and the publisher cannot assume responsibility for the validity of all materials or the consequences of their use. The authors, editors, and the publisher have attempted to trace the copyright holders of all material reproduced in this publication and apologize to copyright holders if permission to publish in this form has not been obtained. If any copyright material has not been acknowledged, please write and let us know so we may rectify in any future reprint.

Trademark Notice: Registered trademark of products or corporate names are used only for explanation and identification without intent to infringe.

Library and Archives Canada Cataloguing in Publication

Handbook of research on food science and technology / edited by Mónica Lizeth Chávez-González, PhD, José Juan Buenrostro-Figueroa, PhD, Cristóbal N. Aguilar, PhD.

Includes bibliographical references and indexes.
Contents: Volume 1. Food technology and chemistry.
Issued in print and electronic formats.
ISBN 978-1-77188-718-2 (v. 1 : hardcover).--ISBN 978-0-429-48785-9 (v. 1 : PDF)

1. Food industry and trade--Technological innovations. 2. Food--Research. 3. Food--Biotechnology. 4. Food--Composition. 5. Functional foods. I. Chávez-González, Mónica Lizeth, 1987-, editor II. Buenrostro-Figueroa, José Juan, 1985-, editor III. Aguilar, Cristóbal Noé, editor

TP370.H36 2018   664'.024   C2018-906012-3   C2018-906013-1

CIP data on file with US Library of Congress

Apple Academic Press also publishes its books in a variety of electronic formats. Some content that appears in print may not be available in electronic format. For information about Apple Academic Press products, visit our website at **www.appleacademicpress.com** and the CRC Press website at **www.crcpress.com**

CONTENTS

About the Editors vii
Contributors ix
Abbreviations xiii
Preface xv

1. **Edible Active Coatings for Foods as a Key Factor for Shelf-Life Prolongation** 1
   Emilio Ochoa-Reyes, Juan M. Tirado-Gallegos, Julio C. Tafolla-Arellano, José Juan Buenrostro-Figueroa, Romeo Rojas, and A. Ochoa-Chantaca

2. **Encapsulation of Prebiotics and Probiotics: A Promising Alternative in Modern Food Technology** 43
   Diana B. Muñiz-Márquez, Jorge E. Wong-Paz, Silvia M. González, Pedro Aguilar-Zárate, and Orlando De La Rosa

3. **Carotenoid Compounds: Properties, Production, and Applications** 63
   Victor Navarro-Macías, Ayerim Hernández-Almanza, Janeth Ventura, Mónica Lizeth Chávez-González, Daniel Boone-Villa, José Luis Martínez, Julio Cesar Montañez, and Cristóbal N. Aguilar

4. **Hydrogels of Biopolymers Functionalized with Bioactive Substances as a Coating for Food Preservation** 89
   Alejandra Isabel Vargas-Segura, Mónica Lizeth Chávez-González, José Luis Martínez-Hernández, Rodolfo Ramos-González, Anna Iliná, and Elda Patricia Segura-Ceniceros

5. **Coffee Pulp as Potential Source of Phenolic Bioactive Compounds** 107
   Luis V. Rodríguez-Durán, Ernesto Favela-Torres, Cristóbal N. Aguilar, and Gerardo Saucedo-Castañeda

6. **Essential Oils in Active Food Packaging** 131
   Olga B. Alvarez-Pérez, Mónica Lizeth, Chávez-González, Anna Iliná, José Luis Martínez-Hernández, Elda Patricia Segura-Ceniceros, Rodolfo Ramos-González, and Cristóbal N. Aguilar

7. **Perspectives for Food Development from Pitayo**
   ***Stenocereus queretaroensis* (Weber) Buxbaum** 149
   Juan José Gaytán-Andrade, Cristóbal N. Aguilar,
   Lluvia Itzel López-López, Luis Enrique Cobos-Puc,
   and Sonia Yesenia Silva-Belmares

8. **Tea from *Camellia sinensis*: A New Tendency of
   Valuable Active Compounds**...................................... 159
   A. Ochoa-Chantaca, G. C. G. Martínez-Ávila, E. J. Sánchez-Alejo,
   and R. Rojas

9. **Polyembryony in Plants and Its Potential in the
   Food Industry** 181
   Ixtaccihuatl Cynthia Gontes-Pérez, José Espinoza-Velázquez,
   Guillermo Cristian Guadalupe Martínez-Ávila,
   Cristóbal N. Aguilar, and Raúl Rodríguez-Herrera

10. **Natural Polymers from Food Industrial Waste as
    Raw Material for Nanostructure Production** 199
    Ariel García Cruz, Rodolfo Ramos-González,
    Mónica Lizeth Chávez-González, Juan A. Ascacio Valdés,
    Arturo I. Martínez, Cristóbal N. Aguilar, José L. Martínez,
    Elda P. Segura-Ceniceros, and Anna Iliná

11. **Arsenic and Heavy Metal Contamination in
    Groundwater in South Asia, Groundwater
    Remediation, and the Vision for the Future**.............. 221
    Sukanchan Palit

Index.. *243*

ABOUT THE EDITORS

Mónica Lizeth Chávez-González, PhD
Full Professor, School of Chemistry of the Universidad Autónoma de Coahuila, Mexico

Mónica Lizeth Chávez-González, PhD, is a Full Professor at the School of Chemistry of the Universidad Autónoma de Coahuila, Mexico, where she develops her work in the Food Research Department. Dr. Chávez-González's experience is in the areas of fermentation processes, microbial biotransformation, enzyme production, valorization of food industrial wastes, extraction of bioactive compounds, and chemical characterization. She is a member of the Sociedad Mexicana de Biotecnología y Bioingeniería and the Asociación Mexicana para la Protección a los Alimentos affiliate of the International Association for Food Protection. She was awarded with the "Juan Antonio de la Fuente" medal for academic excellence and the "Ocelotl" prize for best tecnhological innovation purpose, both given by the Universidad Autónoma de Coahuila. She earned her PhD in Food Science and Technology with an emphasis on valorization of food industrial waste under the tutelage of Dr. Cristóbal N. Aguilar.

José Juan Buenrostro-Figueroa, PhD
Researcher, Research Center for Food and Development, A.C., Mexico

José Juan Buenrostro-Figueroa, PhD, is a Researcher at the Research Center for Food and Development, A.C., Mexico. Dr. Buenrostro has experience in bioprocess development, including microbial processes for enzyme production and recovery of bioactive compounds; valorization of agroindustrial by-products, and extraction and characterization of bioactive compounds from plants and agroindustrial wastes. He has published 17 papers in indexed journals, five book chapters, four patent requests, and more than 45 contributions at scientific meetings. Dr. Buenrostro has been a member of S.N.I. (National System of Researchers, Mexico), the Mexican Society of Biotechnology and Bioengineering (SMBB), and the Mexican Society for Food Protection affiliate of the International Association for Food Protection.

He became a Food Engineer at the Antonio Narro Agrarian Autonomous University. He earned his MSc and PhD degrees in Food Science and Technology from the Autonomous University of Coahuila, México, where he worked on the development of bioprocesses for the valorization of agroindustrial by-products. He also worked at in the Biotechnology Department of the Metropolitan Autonomous University, Mexico City, Mexico.

Cristóbal N. Aguilar, PhD
Full Professor and Dean, School of Chemistry, Universidad Autónoma de Coahuila, Mexico

Cristóbal N. Aguilar, PhD, is a Full Professor and Dean of the School of Chemistry at the Autonomous University of Coahuila, México. Dr. Aguilar has published more than 160 papers published in indexed journals, more than 40 articles in Mexican journals, as well as 16 book chapters, eight Mexican books, four international books, 34 proceedings, and more than 250 contributions in scientific meetings. Professor Aguilar is a member of the National System of Researchers of Mexico (SNI) and has received several prizes and awards, the most important are the National Prize of Research (2010) of the Mexican Academy of Sciences, the "Carlos Casas Campillo 2008" prize of the Mexican Society of Biotechnology and Bioengineering, the National Prize AgroBio-2005, and the Mexican Prize in Food Science and Technology from CONACYT-Coca Cola México in 2003. He is also a member of the Mexican Academy of Science, the International Bioprocessing Association (IFIBiop), and several other scientific societies and associations. Dr. Aguilar has developed more than 21 research projects, including six international exchange projects. He has been advisor of 18 PhD theses, 25 MSc theses, and 50 BSc theses.

He became a Chemist at the Autonomous University of Coahuila, México, and earned his MSc degree in Food Science and Biotechnology at the Autonomous University of Chihuahua, México. His PhD degree in Fermentation Biotechnology was awarded by the Autonomous University of Metropolitana, Mexico. Dr. Aguilar also performed postdoctoral work at the Department of Biotechnology and Molecular Microbiology at Research Institute for Development (IRD) in Marseille, France.

CONTRIBUTORS

Cristóbal N. Aguilar
Autonomous University of Coahuila. Food Research Department, Chemistry Faculty. Blvd. V. Carranza and José Cardenas S/N, Republic East, ZIP 25280, Saltillo, Coahuila, Mexico

Pedro Aguilar-Zárate
Departamento de Ingenierías, Instituto Tecnológico de Ciudad Valles, Tecnológico Nacional de México, Ciudad Valles, 79010, San Luis Potosí, México

Olga B. Alvarez-Pérez
Universidad Autónoma de Coahuila. Food Research Department, School of Chemistry. Blvd. V. Carranza esquina con José Cárdenas Valdés s/n Col, República Oriente, ZIP 25280, Saltillo, Coahuila, México

Daniel Boone-Villa
School of Medicine Unit North, Universidad Autónoma de Coahuila, Piedras Negras, Coahuila, 26090, México

José Juan Buenrostro-Figueroa
Research Center in Food and Development, A.C. 33089, Cd. Delicias, Chihuahua, México

Mónica Lizeth Chávez-González
Universidad Autónoma de Coahuila, Facultad de Ciencias Químicas. Blvd. Venustiano Carranza, Col. República Oriente, C.P. 25000 Saltillo, Coahuila, México,
E-mail: monicachavez@uadec.edu.mx

Luis Enrique Cobos-Puc
School of Chemistry, University of Coahuila, Saltillo, Coahuila, Mexico

Ariel García Cruz
Autonomous University of Coahuila, 25280, Saltillo, Coahuila, México Saltillo, COAH, Mexico

José Espinoza-Velázquez
Agrarian Autonomous University Antonio Narro, Mexican Maize Institute. Calzada Antonio Narro 1923, Buenavista, ZIP 25315, Saltillo, Coahuila, Mexico

Ernesto Favela-Torres
Department of Biotechnology, Autonomous Metropolitan University, Campus Iztapalapa, Av. San Rafael Atlixco 186, Zip Code 09340, Iztapalapa, Mexico City, Mexico

Juan José Gaytán-Andrade
School of Chemistry, University of Coahuila, Saltillo, Coahuila, Mexico,
E-mail: josegaytan@uadec.edu.mx

Ixtaccihuatl Cynthia Gontes-Pérez
Autonomous University of Coahuila. Food Research Department, Chemistry Faculty. Blvd. V. Carranza and José Cardenas S/N, Republic East, ZIP 25280, Saltillo, Coahuila, Mexico

Silvia M. Gonzále
Departamento de Ingeniería Química y Bioquímica. Instituto Tecnológico de Durango, Tecnológico Nacional de México, Durango, 34080, Durango, México

Ayerim Hernández-Almanza
Food Research Department, School of Chemistry, Universidad Autónoma de Coahuila, Saltillo, 25280, Coahuila

Anna Iliná
Autonomous University of Coahuila, Nanobioscience Group, School of Chemistry. Blvd. V. Carranza esquina con José Cárdenas Valdés s/n Col, República Oriente, ZIP 25280, Saltillo, Coahuila, México, E-mail: anna_ilina@hotmail.com

Lluvia Itzel López-López
School of Chemistry, University of Coahuila, Saltillo, Coahuila, Mexico

Arturo I. Martínez
CINVESTAV-Saltillo, Coahuila, Mexico

Guillermo Cristian Guadalupe Martínez-Ávila
Autonomous University of Nuevo Leon, Agronomy School, Francisco I. Madero S/N, Hacienda el Canada. ZIP 66050. General Escobedo, Nuevo Leon, Mexico

José Luis Martínez-Hernández
Universidad Autónoma de Coahuila, Nanobioscience Group, School of Chemistry, Blvd. V. Carranza esquina con José Cárdenas Valdés s/n Col. República Oriente, ZIP 25280, Saltillo, Coahuila, México

Julio Cesar Montañez
Food Research Department, School of Chemistry, Universidad Autónoma de Coahuila, Saltillo, 25280, Coahuila

Diana B. Muñiz-Márquez
Departamento de Ingenierías, Instituto Tecnológico de Ciudad Valles, Tecnológico Nacional de México, Ciudad Valles, 79010, San Luis Potosí, México, Tel.: +52 (481) 38 1 20 44, E-mail: diana.marquez@tecvalles.mx

Victor Navarro-Macías
Food Research Department, School of Chemistry, Universidad Autónoma de Coahuila, Saltillo, 25280, Coahuila

A. Ochoa-Chantaca
Universidad Autónoma de Nuevo León. Research Center and Development for Food Industries (CIDIA). School of Agronomy, General Escobedo, Nuevo León, 66050, México

Emilio Ochoa-Reyes
Centro de Investigación en Alimentación y Desarrollo A.C. 31570, Cd. Cuauhtémoc, Chihuahua, México

Sukanchan Palit
Assistant Professor (Senior Scale), Department of Chemical Engineering, University of Petroleum and Energy Studies, Post-Office-Bidholi via Premnagar, Dehradun – 248007, India, OR 43, Judges Bagan, Post-Office – Haridevpur, Kolkata – 700082, India, Tel.: 0091-8958728093, E-mail: sukanchan68@gmail.com, sukanchan92@gmail.com

Rodolfo Ramos-González
Universidad Autónoma de Coahuila, Facultad de Ciencias Químicas. Blvd. Venustiano Carranza, Col. República Oriente, C.P. 25000 Saltillo, Coahuila, México, E-mail: psegura@uadec.edu.mx

Contributors

Luis V. Rodríguez-Durán
Department of Biotechnology, Autonomous Metropolitan University, Campus Iztapalapa,
Av. San Rafael Atlixco 186, Zip Code 09340, Iztapalapa, Mexico City, Mexico

Raúl Rodríguez-Herrera
Autonomous University of Coahuila. Food Research Department, Chemistry Faculty. Blvd. V. Carranza and Cardenas S/N, Republic East, ZIP 25280, Saltillo, Coahuila, Mexico,
E-mail: raul.rodriguez@uadec.edu.mx

Romeo Rojas
Universidad Autónoma de Nuevo León, Research Center and Development for Food Industries, School of Agronomy, General Escobedo, Nuevo León, 66050, Mexico,
E-mail: romeo.rojasmln@uanl.edu.mx

Orlando de la Rosa
Departamento de Investigación en Alimentos, Facultad de Ciencias Químicas, Universidad Autónoma de Coahuila, Saltillo, 25280, Coahuila, México

E. Sánchez-Alejo
Universidad Autónoma de Nuevo León. Research Center and Development for Food Industries (CIDIA). School of Agronomy, General Escobedo, Nuevo León, 66050, México

Gerardo Saucedo-Castañeda
Department of Biotechnology, Autonomous Metropolitan University, Campus Iztapalapa,
Av. San Rafael Atlixco 186, Zip Code 09340, Iztapalapa, Mexico City, Mexico,
Tel.: +52 55-58 04-4600, Fax: +52 55-580- 4499, E-mail: saucedo@xanum.uam.mx

Elda Patricia Segura-Ceniceros
Autonomous University of Coahuila, 25280, Saltillo, Coahuila, México Saltillo, COAH, Mexico

Sonia Yesenia Silva-Belmares
School of Chemistry, University of Coahuila, Saltillo, Coahuila, Mexico

Julio C. Tafolla-Arellano
Coordinación de Tecnología de Alimentos de Origen Vegetal, Centro de Investigación en Alimentación y Desarrollo, A. C. 83304, Hermosillo, Sonora, México

Juan M. Tirado-Gallegos
Centro de Investigación en Alimentación y Desarrollo A.C. 31570, Cd. Cuauhtémoc, Chihuahua, México, E-mail: jumatirga@hotmail.com

Juan A. Ascacio Valdés
Autonomous University of Coahuila, 25280, Saltillo, Coahuila, México Saltillo, COAH, Mexico

Alejandra Isabel Vargas-Segura
Universidad Autónoma de Coahuila, Facultad de Ciencias Químicas. Blvd. Venustiano Carranza, Col. República Oriente, C.P. 25000 Saltillo, Coahuila, México

Janeth Ventura Sobrevilla
School of Heaalth Sciences, Universidad Autonoma de Coahuila, Piedras Negras, Coahuila 26090

Jorge E. Wong-Paz
Departamento de Ingenierías, Instituto Tecnológico de Ciudad Valles,
Tecnológico Nacional de México, Ciudad Valles, 79010, San Luis Potosí, México

ABBREVIATIONS

AMD	macular degeneration
BF	*Bifidobacterium*
BHT	butylated hydroxytoluene
BSA	bovine serum albumin
C	catechin
CAPE	caffeic acid phenethyl ester
CIDIA	Center and Development for Food Industries
CP	coffee pulp
EC	epicatechin
ECG	epicatechin gallate
EGC	epigallocatechin
EGCG	epigallocatechin gallate
EHEC	enterohemorrhagic *E. coli*
EOs	essential oils
EVOH	ethylene-vinyl alcohol
FDA	Food and Drug Administration
FFC	fresh cut fruits
FOS	fructooligosaccharides
GC	gallocatechin
GCG	gallocatechin gallate
GOS	galactooligosaccharides
GRAS	generally regarded as safe
HPLC	high-performance liquid chromatography
HSCCC	high-speed counter-current chromatography
IPN	interpenetrated network hydrogels
ISAPP	International Scientific Association of Probiotics and Prebiotics
LLDPE	linear low-density polyethylene
LM	low-methoxyl

MAE	microwave-assisted extraction
MAP	modified atmosphere packaging
MRI	magnetic resonance imaging
NPs	nanoparticles
OF	oligofructose
PCL	photochemiluminescence
PHA	polyhydroxyalkanoates
PHWE	pressurized hot water extraction
PLA	polylactic acid
PLGA	poly(lactic-co-glycolic acid)
PP	polypropylene
PTX	paclitaxel
PVA	polyvinyl alcohol
PVAc	acetate to polyvinyl acetate
SSF	solid state fermentation
THF	tetrahydrofuran
TP	tea polyphenols
UC	ulcerative colitis
WVP	water vapor permeability

PREFACE

Human population is growing dramatically, and it will probably reach 10 billion world inhabitants sooner than estimated. It implies important challenges never faced before by humans who need to organize and work on the development of mega-efforts to ensure universal access to health care, food, water, sanitization, energy, education, and housing. These challenges, natural or man-made, obligate the scientific community to proactively seek new breakthrough food and nutrition solutions to ensure global food sustainability and nutrition security in the future. To achieve this, innovative solutions need to be considered throughout the whole food chain, inclusive of food choices and dietary patterns, in order to make significant improvements in the food supply, nutritional, and health status. In the case of foods, innovations in food processing techniques can significantly contribute to meeting the needs of the future world population with respect to quality, quantity, and sustainability of food intake.

Those in academe and industry focused on food science and technology are constantly redefining their traditional forms for new ways to face the threats of the twenty-first century, which is marked by multiple unprecedented environmental challenges that could threaten human survival. The combined impact of climate change, energy and water shortages, environment pollutants, shifting global population demographics, food safety, and growing disease pandemics all place undue stress on the planet's food system, already in a sensitive balance with its ecosystem.

Any changes to the food supply inevitably impact food, nutrition, and health trends and policies, particularly pertaining to food production, agricultural practices, dietary patterns, nutrition, and health guidance and management. As a result, there is an urgent need to find alternative solutions to improve the efficiency and sustainability in the food supply chain by reducing food waste and enhancing nutritional qualities of foods through the addition of nutraceuticals to prepare functional foods and intelligent foods.

The Food Research Group of the School of Chemistry at Universidad Autonoma de Coahuila (DIA-UAdeC) celebrates 25 years of existence and hard work, a period in which it has undergone a tremendous transformation in order to provide solutions and new technological alternatives to the problems demanded by the region, the country, and some international elements. To achieve this, the group grew in the number of researchers, and therefore various lines of research are studied. Today the DIA-UAdeC is formed by research groups in Bioprocesses and Bioproducts, Biorefineries, Biocontrol, Natural Products, Molecular Biology and Ommic Sciences, Glic-Biotechnology, Nano-Bioscience, Edible Coatings, Films and Membrane Technology, Food Engineering, Emerging Processing Technology, Food Science, and Functional Foods.

Two important Mexican postgraduate programs in Food Science and Technology are offered to Mexican and foreign students to whom the National Council of Science and Technology of Mexico (CONACYT) offers scholarships to carry out their MSc or PhD programs.

The consolidation of national scientific cooperation has allowed the prolongation and substantially improvement of the generation and application of knowledge with the Autonomous Metropolitan University, the Autonomous University of Chihuahua, the Autonomous University of Tamaulipas, the Autonomous Agrarian University "Antonio Narro," the Autonomous University of Nuevo León, the University of Colima, the Technological Institute of Durango, the Technological Institute of Ciudad Valles, the Technological Institute of Monterrey, and the centers of research CIQA, CINVESTAV, CIMAV, CIAD, CICATA, CIATEJ, among others.

Strong linkages of international cooperation with institutions and research centers around the world have been established and are now generating important results in the framework of scientific and technological cooperative projects and programs. They highlight their research partnerships with the University of Minho (Portugal); the University of Vigo (Spain); the University of Georgia (USA); the University of Marseille (France); the University of Valle and the National University of Colombia; the National University Nacional de Rosario, the National University of Rio Cuarto and the National University of La Plata (Argentina); Kannur University (India); Federal University of Pernambuco (Brazil); University of Torino (Italy); Jacobs University (Germany); Gachon University (Korea); and other important world-quality research centers including INL (Portugal); IRD and IMBE (France); ICIDCA (Cuba), and the Jawaharlal Nehru Tropical Botanic Garden & Research Institute (India).

For this reason, the research group has organized itself to celebrate its 25th anniversary by publishing a book that reflects the scientific and technological contributions in the field of food Science and technology generated by scientists of the DIA-UAdeC and some of its collaborators.

This *Handbook of Research in Food Science and Technology* consists of three volumes; (i) Food Technology and Chemistry, (ii) Food Biotechnology and Microbiology, and (iii) Functional Foods and Nutraceuticals, all of which will highlight the current trends and knowledge regarding the most recent innovations, emerging technologies, and strategies based on food design on a sustainable level. The handbook includes relevant information on the modernization of food industries, emerging technologies, sustainable packaging, food bioprocesses, food fermentation, food microbiology, functional foods, nutraceuticals, natural products, nano- and micro-technology, healthy product composition, innovative processes/bioprocesses for utilization of by-products, development of novel preservation alternatives, extending the shelf life of fresh products, and alternative processes requiring less energy or water, among other topics.

CHAPTER 1

EDIBLE ACTIVE COATINGS FOR FOODS AS A KEY FACTOR FOR SHELF-LIFE PROLONGATION

EMILIO OCHOA-REYES,[1] JUAN M. TIRADO-GALLEGOS,[1,5]
JULIO C. TAFOLLA-ARELLANO,[2]
JUAN BUENROSTRO-FIGUEROA,[3] ROMEO ROJAS,[4]
and A. OCHOA-CHANTACA[4]

[1] Research Center for Food and Development, A.C. 31570, Cd. Cuauhtémoc, Chihuahua, México

[2] Coordination of Food Technology of Vegetable Origin. Research Center for Food and Development. A. C. 83304, Hermosillo, Sonora, México

[3] Research Center in Food and Development, A.C. 33089, Cd. Delicias, Chihuahua, México

[4] Autonomous University of Nuevo León, Research Center and Development for Food Industries, School of Agronomy, General Escobedo, Nuevo León, 66050, Mexico

[5] Research Center in Food and Development, A.C. Av. Río Conchos s/n, Col. Parque Industrial, C.P. 31570, Cd. Cuauhtémoc, Chihuahua, México, E-mail: jumatirga@hotmail.com

ABSTRACT

In foods, an absence of desirable organoleptic characteristics affects consumer preferences and limits market availability, causing considerable economic losses. Therefore, producers aim to conserve the desirable attributes of food (e.g., texture, color, and flavor). Edible coatings made of polymers generate a modified atmosphere on coated foods. Edible films and coatings can regulate gas exchange, control respiration rate, reduce water loss and decrease the loss of other food components. Moreover, films and coatings can act as carriers of functional ingredients (e.g., antimicrobial and antioxidant agents and vitamins). Thus, active edible polymer coating has become an attractive technology for maintaining food quality and extending shelf life. In this chapter, we provide an overview of recent advances in edible films and coatings from renewable sources.

1.1 INTRODUCTION

The edible films and coatings are used to extend the shelf life of foods preventing loss of some important components (i.e., moisture losses), while selectively allowing for the controlled exchange of important gases involved in respiration processes, such as oxygen, carbon dioxide, and ethylene. Edible films and coatings must meet with a number of specific functional requirements, such as moisture barrier, solute or gas a barrier, water or lipid solubility, color and appearance, mechanical and rheological characteristics and non-toxicity [1, 2].

1.1.1 HISTORICAL ASPECTS

The use of an edible films in food products is not recent, there were natural coverings (such as leaves, wood or baskets) were used long before modern polymers were developed. These natural coverings were used without knowledge of their mechanism. The first record cases of food preservation using a natural covering were in southern China around the twelfth century; the King's citrus fruit was preserved using a molten wax coating [3].

Several centuries later, during the fifteenth century in Japan, a proteinaceous edible film was used to improve the appearance and preservation of food; this is considered the first film coating applied to food [4]. In the 19th century, gelatin was used to preserve meat, which was patented in USA [5]. Sucrose has also been used to coat nuts, almonds, and hazelnuts, thereby preventing rancidity and improving the quality of preservation [5]. In the 1930s, an emulsion of oil-waxes in water was applied on fruits to maintain their bright appearance and to minimize moisture loss during storage. This practice was enhanced by adding various fungicides to prevent rotting [6]. In Europe, this process was known as larding (storing several fruits in wax or fats for later consumption) [7]. The term edible film has been applied to food applications only in the past 60 years. The first patents of an edible films to extend the shelf life of foods date back to the 1960s [8], including films for frozen meat, poultry and sea-food using alginates, fats, gums and starches [9]. These materials have been used for centuries to prevent loss of moisture and to create a shiny fruit surface for aesthetic purposes, and are still used today.

Various food-preservation technologies are currently used (e.g., refrigeration, controlled atmosphere storage, and sterilization by both UV and gamma radiation). Nevertheless, coatings and edible films remain a method to maintain the quality and safety for fresh whole or cut fruit.

1.1.2 COATINGS OR FILMS, WHAT ARE THE DIFFERENCES?

Usually, the terms film and coating are used interchangeably to indicate that the surface of a food has been covered by a relatively thin layer of material. Coatings are applied in liquid form before the thin film is formed directly on the food surface [10]. Although removal of coatings may be possible, they are normally regarded as part of the final product and can be eaten together with foods [11]. The edible film is a thin layer made of an edible material, which, once formed, can be placed on or between food components. Therefore, they are first molded as solid sheets and are then applied as a wrapping on the food product [10]. Edible films are typically less than 0.3 mm [12].

1.1.3 ADVANTAGES AND DISADVANTAGES OF FILMS AND EDIBLE COATINGS

The use of an edible films and coatings can reduce waste accumulation due to the use of non-biodegradable petroleum-based plastics. Edible films and coatings are natural polymers obtained from agricultural productions and are biodegradable, as well as safe for the environment [13].

Edible films and coatings are involved in the control of mass transfer between the food and its environment, as well as between food components.

This might extend the shelf life and maintain or improve the condition of packaged food [14]. Most of the time, natural biopolymer-based films have limited mechanical and barrier properties compared to the most common synthetic plastics used as packaging in the food industry [15]. However, the barrier properties of an edible films and coatings might be compromised by poor mechanical properties, resulting in failure to maintain the film integrity during handling, packaging, and carrying processes [16].

According to Guilbert et al. [17] biopolymer-based films primarily composed of polysaccharides or proteins, have suitable mechanical and optical properties, but are highly sensitive to moisture and show poor water vapor barrier properties. In contrast, films composed of lipids or polyesters have good water vapor barrier properties, but are usually opaque, relatively inflexible, quite fragile and unstable (rancidity). Edible coatings are promising systems for the improvement of food quality preservation during processing and storage. Moreover, these materials could be used as an alternative to polymeric plastics.

1.1.4 FILM AND COATING FORMULATION

There are no basic differences in the material composition of films and coatings. Usually, both can be obtained from the same formulation. Normally, films and coatings are designed using biological materials such as polysaccharides, polyester proteins, lipids, and derivatives. The edible films are produced exclusively from renewable, biodegradable, and edible ingredients, which degrade more rapidly than polymers derived from petroleum.

Edible films and coatings can be produced from animal or plant proteins [18, 19]. Proteins are macromolecules with specific amino acid sequences and molecular structures and are commonly used as film-forming materials. The most distinctive characteristics of proteins (compared to other film-forming materials) include their confirmational denaturation, electrostatic charges, and amphiphilic nature. The secondary, tertiary, and quaternary structures of proteins can be easily modified to achieve the desired film properties (by the use of heat, pressure, irradiation, mechanical treatment, acids, alkalis, metal ions, salts, chemical hydrolysis, enzymatic treatment, and chemical cross-linking) [20].

Polysaccharide film-forming materials include starch, pectin, gums, and fibers. Compared with proteins, the confirmation of polysaccharide structures is more complicated and unpredictable, resulting in larger molecular weights than encountered with proteins. Most carbohydrates are neutral, while some gums are charged negatively or (rarely) positively. Thus, the large numbers of hydroxyl groups or other hydrophilic moieties in the neutral carbohydrate structure play a significant role in film formation and characteristics. Some negatively charged gums, such as alginate, pectin, and carboxymethyl-cellulose, show different rheological properties in acidic conditions compared to neutral or alkaline conditions, as well as in the presence of multivalent cations [20]. The film-forming ability of several polysaccharides has been studied, including cellulose, chitosan, starch, pectin, alginate, carrageenan and pullulan [21–24].

Lipid-based materials include waxes, acylglycerols, and fatty acids [25]. Although, the lipids and resins are used as film-forming materials, they are not polymers. However, they are edible, biodegradable, and cohesive biomaterials. Most lipids and edible resins are soft solids at room temperature and have characteristic phase transition temperatures. Lipids can be manipulated to any shape by casting and molding systems after heat treatment, causing reversible phase transitions between fluid, soft-solid, and crystalline-solid states [20].

In most cases, plasticizer components are required to produce edible films and coatings, especially when using polysaccharides and proteins. Plasticizers are small-molecular-weight compounds that can be added to an edible film or coating solution to improve the flexibility and handling, maintaining integrity and reducing the number of pores and cracks in the polymeric matrix that can be formed during the coating process or during frying [26]. Also, the plasticizers increase the free volume of polymer structures or the molecular mobility of polymer molecules, which affect the mechanical properties and resistance of an edible films and coatings to permeation of vapors and gases [13, 27]. The use of natural plasticizers, such triglycerides from vegetable oils or fatty acid esters, is increasing because of their low toxicity and low migration properties [28].

Antioxidants, antimicrobials, nutraceuticals, flavors, and colorants can also be incorporated into film-forming solutions to achieve active packaging or coating functions. Adding antioxidants and antimicrobial agents into films provides additional active functions (e.g., protecting food products from oxidation and microbial spoilage), resulting in quality improvement and enhanced safety [29, 30]. Incorporated flavors and colorants can improve the organoleptic preference and the visual perception of quality [20].

The combination of polymers to form films could involve proteins and carbohydrates, proteins and lipids, carbohydrates and lipids or synthetic polymers and natural polymers. The main objective of producing composite films is to improve the permeability or mechanical properties of the film [31]. Kester and Fennema [32] introduced emulsion films from methylcellulose and fatty acids to improve the water vapor barrier function of cellulose films. In general, all of the components used in edible films and coatings must have the GRAS (generally recognized as safe) status.

1.2 EDIBLE FILMS AND COATINGS FROM NATURAL MATERIALS

Food packaging is used to achieve aesthetic, barrier, and mechanical. An ideal packaging would be produced from inexpensive and renewable raw materials. Over the years, several packaging materials have been used, incorporating diverse bioactive components into the film or coating (e.g., polysaccharides, proteins, lipids, waxes, and resins) [33]. Efforts have been made to improve the maintenance of quality, freshness, and food safety, which has included the use of sustainable material in food packaging [34]. Also, world production of plastic resins has increased around 25-fold, while less than 5% of all plastics are recycled, leading to a rapid accumulation of plastics in the environment [35]. As a result, the research focus has been directed to biodegradable biopolymers as an alternative to plastics.

1.2.1 EDIBLE FILMS AND COATINGS USING POLYSACCHARIDES

Polysaccharides, including pectin [22], starch [36], cellulose [37], chitin [21, 38], and Arabic gum [39], have used to formulate edible films and coatings. Polysaccharides can be used as an alternative or replacement of traditional materials, thereby reducing traditional polymeric packaging. Edibility and biodegradability are the most beneficial characteristics of an edible films and coatings [40]. Also, plasticizers and others food grade additives have been used to improve the characteristics of films and coatings. Table 1.1 shows the most recent design, application and effect of the use of an edible films and coatings on several products.

1.2.2 EDIBLE FILMS AND COATINGS USING PROTEINS

The use of proteins to formulate edible films and coatings has been widely used to extend the postharvest life of fresh fruits, vegetables, and other products. Edible films and coatings can provide a substitute for modified atmospheric packaging [48]. Recently, edible films from multiple protein sources have been formulated, including casein [49], gelatin [50, 51], whey protein [52, 53], corn zein [54, 55], soy protein [56, 57], wheat gluten [58, 59], peanut protein [60, 61], and mug bean protein [62]. The low relative humidity of protein films acts a good oxygen barrier and provides nutritional value and good mechanical resistance, as well as providing an excellent barrier to aroma compounds and oils [52]. In addition, to formulate edible films and coatings it is necessary the use of plasticizers to improve their flexibility and reduce the cohesion within the film network. Plasticizers do this by entering between polymeric molecular chains, which are physically and chemically associated with the polymer.

1.2.2.1 ANIMAL PROTEINS

Several animal proteins are used to formulate edible films and coatings. The proteins most used are gelatin [51, 63], whey protein [64, 65], collagen [66, 67], egg protein [68], shrimp muscle proteins [69], squid mantle muscle [70] and bone meal protein [71]. Some of these proteins are sourced from the waste present in the bones and skins generated during animal slaughtering and processing. Also, the unique ability of proteins to form networks and induce plasticity and elasticity is considered beneficial in the preparation of biopolymer-based packaging materials [72]. However, fish gelatin has gained importance in recent years because: (1) the outbreak of bovine spongiform encephalopathy during the second half of the 1990s; (2) the banning on consumption of collagen from pig skin and bone in some religions (e.g., Halal and Kosher); and (3) some technological advantages over mammalian gelatins [73]. In addition, bioactive

TABLE 1.1 Edible Films and Coating Based on Polysaccharides

Product	Components	Variables	Main results	Reference
Red Crimson Grapes	Waxy-based cornstarch and gelatin, plasticized with GLY or SOR.	WS, SE, (C, B, A, F; T)	Less WS than the control group. Sensory evaluation showed that all coating did not affect acceptability scores.	[41]
Fresh-cut apples	CMCS (8 mg/mL, pH = 7.2), CS (4 mg)	WS ratio, Firmness	PEM coating performed well on WL inhibition, maintaining fruits firmness and preserving apple quality.	[42]
Beef Steaks	CS (0.5–1.5%), Gelatin (0–6%), GLY based on dry gelatin (0–12%)	WS analysis, LO	The application of gelatin-, chitosan- and glycerol-based biopolymers effectively decreased the weight loss, lipid oxidation and discoloration of coated steaks during 5 days of retail.	[43]
Tomatoes	CS, I, GAA, Catechin	WS, AA, TPC	Extends the shelf life of tomatoes without any change in antioxidant activity.	[44]
Fresh-cut Pineapples	CS, PU, LM, NM, AM	PCP, M, GPM, SE	Improved the quality and prolonged the useful life the freshly cut pineapple for 6 days compared to the control.	[45]
Breba figs	Whit a mucilage solution of Opuntia ficus-indica cladodes	SE, SS, pH, TA, TPC, C	The coating showed a tendency to extend the breba figs shelf life reduced the development of Enterobacteriaceae.	[46]
Kiwifruit slices	Mucilage	FN, WS, SE, ASA, M, PA	Low WS; higher firmness until 5 days of shelf life.	[47]

WS: weight loss, **M:** microbiology; **TPC:** total phenolic content, **C:** carotenoids, **AA:** antioxidant activity, **TP:** texture properties, **SE:** sensory evaluation, **TexP:** texture profile, **SOR:** sorbitol, **GLY:** glycerol, **C:** color, **B:** brightness, **A:** aroma, **F:** flavor, **T:** texture, **CMCS:** carboxymethyl-cellulose sodium, **CS:** chitosan, **PEM:** polyelectrolyte multilayer, **LO:** lipid oxidation, **I:** iodide, **GAA:** glacial acetic acid, **PU:** pullulan, **LM:** linseed, **NM:** nopal cactus, **AM:** aloe mucilage, **FCP:** physicochemical parameters; **GPM:** growth of pathogenic microorganisms, **SS:** soluble solids, **TA:** titratable acidity, **ASA:** ascorbic acid, **PA:** pectin analysis, **FN:** firmness.

compounds, such as antioxidants and antimicrobial agents have been added to improve the shelf life of several products [51]. This is advantageous because these products can be consumed by people intolerant to allergens such as gluten (celiac), as well as having increased nutritional value. Table 1.2 shows the most recent (last 5 years) studies addressing the formulation and application on several food products.

1.2.2.2 VEGETABLE PROTEINS

Vegetable proteins are also used to formulate edible films and coatings, including gluten [58, 79, 80], corn protein (zein) [54, 81], hazelnut meal protein [82], bitter vetch protein [83, 84], soy protein [85], quinoa protein [86], etc. This trend has been driven by an increase in global economic problems, as well as consumers demand originated from health concerns, religious limitations, vegetarianism, and interest in functional plants as an alternative to animal proteins [82, 87]. Moreover, vegetable proteins are a potential candidate material for developing edible films and coatings, such as a biodegradable alternative for packaging materials. Also, the fast grow of plants is advantageous. Protein-based edible films and coatings have effective gas-barrier and mechanical properties compared with those from lipids and polysaccharides [88], as well as traditional low-density polyethylene films [89]. These properties create an opportunity for low-cost food preservation, including nutritional content, good biodegradability and biocompatibility. Table 1.3 summarizes recent advances in the use of an edible films and coatings based on vegetable proteins.

1.2.3 EDIBLE FILMS AND COATINGS USING LIPIDS, WAXES, AND RESINS

Each of the components of an edible film and coating has a specific function. The polymer is the base matrix; whereas plasticizers are added to the solution to enhance softness, flexibility, elongation, and clarity. Waxes, lipids, and resins are added to provide an additional moisture barrier due to their low affinity for water [93]. Lipids, waxes, and resins have been used for centuries to prevent moisture loss and create a shiny product surface. These include candelilla [94–96], carnauba [97, 98], shellac [99, 100], bees

TABLE 1.2 Edible Films and Coating Based on Animal Proteins

Product	Components	Variables	Main results	Reference
Chocolate	H, P, and CB and plasticized with sucrose.	WVP, SE	The films obtained were easily manageable and flexible.	[66]
Mangoes Apples	Galactomannans, Collagen and GLY	W, WVP	That these coatings can reduce gas transfer rates in these fruits, and can be extend their shelf life	[74]
Apples	EC	GC/MS analysis, Gas Exchange measurements.	Acetate esters, normally increasing with maturity, were less concentrated in coated apples (−78% 2-methylbutyl acetate and −73% hexyl acetate, after 1 and 7 days respectively)	[75]
Cooked meatballs	WP, antioxidant (LA, S)	TPC, PV, SE, TAV	Total phenolic compound content was higher in meatballs treated with NAE-WPF than in the C group suggesting that these natural compounds contributed to the oxidative stability of frozen meatballs	[53]
Smoked Salmon	WP, LYS	LYS activity, MP, MC	Immobilization of LYS (+) was shown by the films of the pH ≥5.5. The release of LYS was initiated when the pH was reduced below 5.5.	[76]
Fish Sausage	CS, Lactic acid, Gelatin, GLY	ANA; WC; WHC; SE	The coating increased the delay phase of total viable microorganisms and enterobacteria at 15 and 10 days, respectively, while the film drastically inhibited the growth of these groups	[77]
Smoked Salmon	CFP, Gelatin, SOR	FC, FT, TS, EB, MC, WSol, WVP, M,	The tensile strength and elongation at break of the CFP/gelatin composite film significantly ($p < 0.05$) increased as the gelatin content in the film increased	[78]

HC: hydrolyzed collagen, CB: cocoa butter, WVP: water vapor permeability, SE: sensory evaluation, P: Pectin, GLY: glycerol, W: wettability, EC: edible coating, GC: Malus x domestica Borkh, MS: CO2 emission, LA: laurel, S: sage, C: control, TPC: total phenolic content, PV: peroxide value, TAV: thiobarbituric acid value, WP: whey protein, LYS: lysozyme, MP: mechanical properties, MC: moisture contents, (+): positive charge, (−): negative charge, CS: chitosan, GLY: glycerol, ANA: antimicrobial activity, WC: water content, WHC: water holding capacity, CFP: chicken feather protein, SOR: Sorbitol, FC: film conditioning, FT: film thickness, TS: tensile strength, EB: elongation at break, WSol.: water solubility, WVP: water vapor transmission.

TABLE 1.3 Edible Films and Coating Based on Vegetable Proteins

Product	Components	Variables	Main results	Reference
Cheese	Biodegradable Zein	CA, TPA	The samples with the biodegradable coating presented 30% less WL and microbiological contamination during 50 days.	[90]
Strawberries	CS, Q, SO	SE, FN, WS	Fresh strawberries coated with CS, Q/CS/SO and Q/CS coatings had longer shelf lives than uncoated fruits. This effect is mainly due to the antifungal activity of CS, which remains when CS is combined with Q and OS.	[91]
Fresh-cut apples	Soy protein, Ferulic acid	WS, CA, FN, pH, SS	The incorporation of ferulic acid at a concentration of 4.0 g L^{-1} in the soy protein film-forming solution (30 g L^{-1}), at pH 7.0, sufficiently away from the isoelectric point of protein, decreases the WVP of the soy protein film. This result can be attributed to the ferulic acid cross-linking properties with soy protein and its hydrophobic character.	[57]
Toasted groundnut	Cassava starch and soy protein concentrate (100:0 and 50:50). 20% Glycerol	Moisture content, Sensory evaluation.	Extending the shelf life of toasted groundnut during ambient (27 ± 1°C) storage for 14 days. The use of an edible coatings maintained the shelf life and preserve the sensory parameters.	[92]

SPI: Soy protein isolate, GLY: Glycerol, APLQ: Assessment of Persian lime quality, MACPLS: Microstructural analysis of coating on Persian lime surface, TPA: Texture profile analysis, CA: Color analysis, WS: Weight loss, CS: Chitosan, Q: Quinoa, SO: Sunflower oil, SE: Sensory evaluation, FN: Firmness, SS: Soluble solids, WVP: Water vapor permeability.

TABLE 1.4 Edible Films and Coating Based on Lipids, Waxes, and Resins

Product	Components	Variables	Main results	Reference
Strawberries	*Candelilla* wax, *Bacillus Subtilis* HFC103	Application of the edible coating	The Bacteria treatment was the most effective to control *R. stolonifer*. However, it did not show the same efficacy to control weight loss or to prolong shelf life of strawberry. Results showed that the combination of candelilla wax edible films and the inoculation with *B. subtilis* HFC103 is an innovative strategy with the potential to prolong shelf life of strawberry.	[95]
Banana	Shellac, Gelatin	pH, TA, ST, ASC, FN, WS	The treatment was effective to retard the maturation process and reductions were obtained in all parameters, such as weight, texture, acid, sugar and total yeast/mold counts.	[99]
Eggplant	CW, PEG, SA, SDS	ASP, PWL, FN, PA, TPC, EAC	Effectiveness of additives such as PEG, SA and SDS in CW emulsion coating could extent shelf life and enhanced antioxidant activity of packaged eggplants at maximum storage period during ambient temperature.	[113]
Guava	AG, SC, TE	SQA, O, CD, WVT	The AG, SC, and TE registered the significant effect of altering the O_2, CO_2, and water vapor transmission rate of the coated guava. The optimized coating formulation was 5 g/100 mL AG, 1 g/100 mL SC and 2.5 mL/100 mL tulsi extract	[39]
Plum	HM, OEO, Bergamot	SE, QPP, AEP, QP	The 2% H-OEO coating was shown to be very effective in reducing the respiration rate, ethylene production, total weight loss, and total cell count	[114]
Low-fat cut cheese	NE, OEO	V, WS, WVP, TPA	Edible coatings with at least 2.0% w/w of OEO improved the microbial stability of the cheese pieces, resulted effective in the decontamination of external pathogens such as Staphylococcus aureus and preserved cheese outward appearance during the time.	[115]
Strawberries Apples	CS, OOR	WS, SS, TA, FS	Increased antifungal and antibacterial activities. The integration of OOR into CS can be used to improve the inhibition properties of CS based film against spoilage and pathogenic strains	[116]

WS: Weight loss, TA: Titratable activity, ST: Surface tension, ASC: Adhesion and spreading coefficient, CW: Carnauba wax, ASP: Analysis of storage properties, PLW: Physiological loss in weight, FN: Firmness, PA: Peroxidase activity, TPC: Total phenolic content, EAC: Equivalent antioxidant capacity, PEG: Poly ethylene glycol, SA: Sodium alginate, SDS: Sodium dodecyl sulfate, AG: Arabic gum, SC: Sodium caseinate, TE: Tulsi extract, SQA: Subjective quality analysis, O, oxygen, CD: Carbon dioxide, WVT: Water Vapor transmission, STH: Starch, GLY: Glycerol, LP: Lipid, ANT: Antioxidant, M: Microbiology, FT: Film thickness, FS: Film solubility, FTP: Film transparency, AT: Analytical test, OEO: Oregano essential oil, NE: Nanoemulsion, V: Viscosity, WVP: Water vapor permeability, TPA: Texture profile analysis, HM: Hidroxypropyl Methylcellulose, SE: Sensory evaluation, QPP: Quality parameters of plum, AEP: Antimicrobial evaluation for plums, QP: Quality parameters of the plums, CS: Chitosan, SS: Soluble solids, OOR: Olive oil residues extract.

[101–103], rice [104], oregano essential oil [105], palm fruit oil [106], lime essential oil [107], lemongrass essential oil [108], olive oil [109], gum Arabic [110], gum tragacanth [111], mesquite gum [106], karaya gum [93], ghatti gum [112], vainilla oleoresin [93], and mastic resin [93]. Table 1.4 shows the most recent advances in the use of lipids, waxes, and resins to formulate and evaluate edible films and coatings.

1.3 EXTENDING SHELF LIFE OF VEGETABLES AND FRUITS WITH EDIBLE FILMS AND COATINGS

Fruits and vegetables are an important part of the human diet. These foods are rich in nutrients and phytochemicals, which are important for good health. Most fruits and vegetables are perishable and have a limited shelf life due the postharvest desiccation and pathogen attack, thus limiting their commercialization (both in local and international markets). The cuticle is a protective natural wax layer covering fruits and vegetables, which is involved in regulating water loss and gas exchange, as well as biotic and abiotic stresses [117]. However, once fruits are processed or handled, their natural protection can be damaged, and this could be avoided by the application of coatings [118]. An edible coating is a thin film prepared from edible material that acts as a protective barrier of external elements, thereby prolonging the product's shelf life [119].

1.3.1 PHYSIOLOGY OF FRESH AND MINIMALLY PROCESSED FRUITS AND VEGETABLES

Fruits and vegetables are living tissues and their physiological process remains active after harvest. Therefore, to successfully extend their shelf life, it is important to understand the physiology of fresh and minimally processed fruits and vegetables. Mainly, efforts are related to vital metabolic processes that keep the fruits and vegetables alive until consumption. For instance, respiration is influenced by several factors, such as stage of development, environment conditions and some stress associated with the minimal processing. It has been reported that minimally processed fruit and vegetables have higher respiration rates compared to intact fruit and vegetables, and this triggers the rapid deterioration of acids, sugars, and nutrients. Also, enzymatic activity leads to browning, affecting the appearance and shelf life [120, 121]. Moreover, the high-water content (80–90%) of fruits and vegetables influences their shelf life by water loss, softening, and, consequently, pathogen attack [122]. Several technologies have been used to control the metabolic processes involving respiration and enzymatic activities, such as low temperatures and modified atmospheres. Otherwise, the applications of films and coatings have been effective at controlling these processes and shelf life prolongation.

1.3.2 APPLICATION OF FILMS AND COATINGS ON FRUITS AND VEGETABLES

The application of an edible films and coatings to fresh fruits and vegetables allows the regulation of respiration and senescence in a similar way to modified atmospheres. Thus, these materials behave as a barrier to gases and the water vapor, thereby slowing the deterioration of fruits and vegetables [123]. According to Olivas and Barbosa-Cánovas [124], relevant questions that should be considered when formulating coatings include:

a) Could the properties of the solution affect the product?
b) What is the stability of the coating during the storing time?
c) How will the film behave under the storage conditions? and
d) What are the desired physicochemical (thickness, color, transparency, solubility), barrier properties, and taste of the coating?

All of these issues are strongly influenced by the coating solution composition [123]. In general, edible films and coatings consist of polysaccharides, proteins, hydrophobic materials (lipids and waxes) and additives (e.g., plasticizers). All of these materials can be used alone or in combination (composites) [125–127]. Lipids and waxes have long been used as edible films/coatings for the conservation of fresh products. Lipid-based edible films and coatings act as a barrier against water vapor and confer brightness to the surface of the product. Most of these materials are applied in form of emulsion with other components [10, 128]. Polysaccharides are the most commonly used natural polymers in the manufacture of an edible films and coatings for fruits and vegetables [129]. These polysaccharides, particularly starches, have low oxygen permeability, which decreases the respiration rates of fruits and vegetables [130]. Starch coatings have been used to retard strawberry fruit senescence [131, 132]. Starch has also been modified and used in combination with other polysaccharides to design novel coatings. Kittur et al. [133] evaluated the effect of different composite coating formulations based on polysaccharides (carboxymethyl-cellulose, carboxymethyl starch, hydroxymethyl starch, hydroxypropyl starch, and N,O-carboxymethyl chitosan) to maintaining quality and lengthen the shelf life of banana and mango at 27 ± 2°C. They found that starch and cellulose-based coatings reduced respiration rates, however, the weight loss was not significant. On the other hand, chitosan-based coatings significantly reduced the weight loss and respiration rate compared to the control. Moreover, chitosan-based coatings also protected against fungal infections. Among the polysaccharides used in formulation of an edible films and coatings applied to vegetables, chitosan showed promissory results because of its antimicrobial activity. Also, chitosan is a cationic polysaccharide and can interact with other anionic components [134]. Proteins are another biopolymer used in coating formulations, providing acceptable mechanical properties, and being good barriers to gases diffusion but poor water vapor barriers [133]. Animal proteins used to make edible coatings and films include caseins, whey protein, collagen, and gelatin keratin, whereas proteins of plant origin include wheat gluten, soy protein, peanut protein, corn-zein, cotton-seed protein, among others [122]. Protein-based edible films and coatings have been used to lengthen the shelf life of fruits and vegetables [41, 133, 135–136]. Most edible film/coating formulations include plasticizer agents to enhance their mechanical properties. However, plasticizers such as glycerol and other low-molecular-weight polyhydroxy-compounds increases the water vapor permeability (WVP) of films and coatings [122]. This is inconvenient because vegetable products are living tissues with high water content and, therefore, the WVP of the coatings must be low so as to slow the senescence process during storage. Therefore, some hydrophobic components (e.g., lipid or waxes) have been added to edible film and coating formulations. Therefore, Vargas et al. [137] formulated a coating based on high-molecular-weight chitosan combined with acid oleic to preserve the quality of strawberries under cold storage. The authors observed that the addition of oleic acid improved the antimicrobial activity of chitosan and the WVP resistance of chitosan-coated samples. In other cases, the polymers are used in combinations. Abugoch et al. [138] evaluated the effect of coatings composed of quinoa protein/chitosan/sunflower oil for extending the shelf life of fresh blueberries stored at 4°C and 75% relative humidity. These authors found that the shelf life of fresh blueberries could be lengthened by the coating and that this was because of reduction in mold growth during storage.

1.3.3 APPLICATION OF FILMS AND COATINGS ON MINIMALLY PROCESSED FRUITS AND VEGETABLES

Lifestyle and dietary habits influence the food consumption trends and ready-to-eat, minimally processed fruits and vegetables with a natural appearance and healthy properties are now in high demand. Minimally processed fruits and vegetables are also referred to as fresh-processed, lightly processed, partially processed, pre-prepared, ready-to-use [121, 139]. These terms describe any fruit or vegetable that has been physically altered from its original form but maintaining similar characteristics to the whole original fruit or vegetable. This might include packaging (including washing), peeling, slicing and disinfection [120, 140]. However, these processes trigger physiological and biochemical changes, such as water loss, increased respiration, ethylene production and microbial contamination, which alter the quality and shelf life of minimally processed fruits and vegetables [141]. A minimally processed product with good quality should be characterized by a fresh appearance, acceptable texture, good taste and odor, microbiological safety, and sufficient shelf life to allow the product to be included within a distribution system [141, 142]. The technologies used to address this problem can be classified into three categories: physical-based preservation, chemical-based preservation and biopreservation technology [142]. Moreover, the development of novel packaging materials, such as edible films and coatings, are new trends in food packing. Films and coatings have contributed to lengthening the shelf life of minimally processed fruits and vegetables by regulating the water and gas exchange as well as microbial spoilage [123, 140, 142]. Table 1.5 gives examples of the composition, effect, and method of application of films and coatings on minimally processed fruits and vegetables.

1.3.4 ACTIVE PACKAGING IN FRUITS AND VEGETABLES

The main goal of using edible films and coatings in fruits and vegetables is to provide a semipermeable barrier against the diffusion of gases and water vapor [123, 143]. Furthermore, the materials of films and edible coatings can function as carriers for the controlled release of additives that might prolong shelf life [126]. To this end, edible coating formulation might include antimicrobial agents, antioxidants, nutrients, and flavors [118, 127, 143, 144]. Some active edible films and coatings used to extend the shell-life of fresh and minimally processed fruits and vegetables are shown in Table 1.5.

1.3.5 METHODS OF COATING APPLICATION ON FRUIT AND VEGETABLES

The application method has an important role in the use of coatings in foods. The most common methods employed are spraying, dipping, and dripping [118, 122, 127]. To achieve its proper purpose, the product must be fully covered by the coating and dried as soon as possible [118]. When scaling-up from the laboratory to industrial levels, it is important to consider that these methods are strongly influenced by the coating formulation. And that an efficient drying system is required [118, 143]. Table 1.5 summarizes the various coatings applied to fresh and minimally processed fruits and vegetables, as well as their application methods.

1.4 EXTENDING SHELF LIFE OF MUSCLE FOODS WITH EDIBLE FILMS AND COATINGS

Edible films and coating enhance the quality of foods, controlling the mass transfer, moisture, oil diffusion, gas permeability (CO_2 and O_2), flavor and aroma losses, thereby maintaining the mechanical and rheological characteristics, color, and appearance in foods, and ultimately lengthening the food' shelf life. Also, edible films and coatings protect the foods from biological deterioration, light-induced chemical changes, and oxidation of nutrients [151]. Edible films and coatings are considered biodegradable packaging due to use of biological materials such as lipids, proteins,

TABLE 1.5 Active Edible Coatings Applied in Fresh and Minimally Processed Fruits and Vegetables

Fruit/vegetable state	Coating formulation and method of application	Main results	Reference
FFV Tomatoes	STC+GLY+CCO+GTE MOP: Dipping	Showed a reduced weight loss coated tomatoes as compared to control. Additionally, coconut oil and tea leaf extract in the edible coating exhibited delayed ripening effects and microbe-barrier properties on tomatoes.	[145]
Arbutus unedo fruit	ALG+EU/CIT MOP: Dipping	The edible coatings of AL were the best to maintain most quality attributes. The incorporation of Cit and Eug into the alginate edible coatings improved the coatings in most cases, preserving sensory and nutritional attributes and reduced microbial spoilage.	[146]
Persian lime	SPI+CIT 10%/LMN% MOP: Dipping	Postharvest protection analyses demonstrated the effectiveness of the SPI-coating as a carrier of antimicrobial compounds against blue mold and reducing water losses as well as maintain color was observed	[147]
FVMP Apple slices	kC/WP/CMCS/PEG/ASA/AC/ CaCl$_2$ MOP: Dipping	Prolonged the shelf life up to two weeks. Antibrowning and CaCl$_2$ help to maintaining color during storage and reduce the firmness loss, respectively.	[148]
Papaya cubes	βCD/TCA+CH+P MOP: Dipping	Shelf life of uncoated fruits was <7 days at 4°, while antimicrobial edible coatings extended the shelf life of fresh-cut papaya up to 15 at 4°C. Coatings reduced the loss of vitamin C and carotenoids	[149]
Apples Potatoes Carrots	WP/P+TG, SPI+GLY+CaCl$_2$ MOP: Dipping	Coatings reduced weight loss of samples coated about 20 and 40%, respectively. When transglutaminase was incorporated in whey protein-pectic coating, weight loss observed was about 80%.	[150]

FFV: Fresh fruits and vegetables, FVMP: Fruits and vegetables minimally processed, SPI: Soy protein, EU: Eugenol, CIT: Citral, LMN: Limonene, STC: Starch, GTE: Green tea extract, CCO: Coconut oil, CH: Chitosan, Pectin: P, GLY: Glycerol, WP: Whey protein, TG: Transglutaminase, MOP: Method of application, kC: k-Carrageenan gum, PEG: Polyethylene glycol, ASA: Ascorbic acid, CA: Citric acid, CMCS: Carboxymethyl-cellulose sodium, βCD: Beta-cyclodextrin, TCA: Trans-cinnamaldehyde.

and polysaccharides employment in their production; which can be easily degraded by the action of bacteria, fungi, and algae [152]. Edible films and coating from proteins are preferred to polysaccharides due to their ability to form films with better mechanical and barrier properties. However, these properties depend on the type of protein, their source, structure and amino acid profile. Among the proteins used in the formulation of an edible film and coating, collagen and gelatin are highlighted for their film-forming ability, which makes them widely used [153].

1.4.1 COLLAGEN AND GELATIN COATINGS

Collagen, an acid-soluble protein, is the major fraction of connective tissue in animals, and is used to make coatings because of its oxygen barrier properties. The partial hydrolyzation or denaturation by a thermal process produces a gelatin formation [152]. Gelatin reduces oxygen and moisture migration and improves the texture and increases the water-holding capacity of food products [154]. Typical sources of collagen and gelatin include bovine hides, bones, and pigskins. Edible films and coatings with collagen and gelatin are widely used in medical, pharmaceutical, and cosmetic applications. To improve the mechanical properties, gas exchange, and adherence to surfaces, two or more components should be combined. A plasticizer (such as glycerol, sorbitol or polyethylene glycol) could be added to modify the mechanical properties of the film and increase its flexibility, workability or distensibility [28]. Benzoic acid was added to a gelatin coating formulation applied to tilapia [155]. This study found that microbial growth was reduced and sensory characteristics of the tilapia were maintained by the coating.

According to Benjakul et al. [153] intrinsic factors (amino acid composition, molecular weight distribution and sources of film-forming material), processing parameters (heat treatment and pH adjustment), film drying condition (drying rate and environmental conditions), and some additives and protein properties (use of plasticizer, cross-linker and protein modifier) affect the properties of films and coatings. For example, oregano and rosemary extracts were used as bioactive compounds with different sources of gelatin in an edible coating [156]. The mechanical properties, water solubility, and water vapor permeability did not show changes, except using tuna skin gelatin, which decreased the deformability and increases the water solubility of the product.

1.4.2 APPLICATION IN MEAT PRODUCTS

Meat is defined as the flesh of an animal used as food. Meat and meat products are highly perishable, and their shelf-lives depends on the animal, including differences between the breed and muscle fiber type. Some external factors, such as diet and stress, affect the shelf life of meat products before the animal is sacrificed. Once sacrificed, post-mortem storage conditions (time, temperature, and packaging atmosphere) affect the length of time that the meat product will be acceptable for consumers [154]. Slight changes in specific characteristics (purge accumulation, unacceptable aroma, appearance, color deterioration, rancidity and microbial spoilage) indicate to the consumer that the meat is old or damaged. Also, there are intrinsic factors that accelerate the spoilage of fresh meat and products, such as water activity, pH, and endogenous enzymes. Meat typically has a water activity value greater than 0.85, which, combined with a favorable pH, allows the growth of a broad range of spoilage microbes. This microbial growth results in several changes in meat products, including color and aroma deterioration [157]. The use of an edible films and coatings containing gelatin and collagen in meat products has been reported to reduce purge, preserve color, decrease the aroma deterioration, improve sensory scores, slow spoilage, and act as an antioxidant [154].

Antoniewski et al. [158] evaluated the effect of a gelatin coating on the shelf life of fresh meat. They found that a 20% bovine gelatin solution reduced purge by acting as a barrier to water loss. Also, reduced color deterioration was observed in beef and pork, but not for salmon and chicken, when this was applied.

1.4.3 APPLICATION IN FISH PRODUCTS

Fish is an excellent source of quality proteins and other nutrients for the human diet, and fresh consumption is greatly preferred over frozen. However, fresh fish has a high-water content and large quantity of free amino acids and volatile nitrogenous bases. These properties make fish highly susceptible to chemical and biological deterioration (e.g., such as enzyme and microbial degradation). Chemical and enzymatic reactions are the main cause of initial deterioration, followed by microbial spoilage, which reduces the shelf life [159].

Much research has focused on developing novel methods and technologies for the preservation of fresh fish products, including numerous films and coatings. A mixture of gelatin and lignin reduced the lipid oxidation of salmon fillets submitted to high pressure and stored in refrigeration [160]. Volpe et al. [161] studied the effect of a carrageenan coating with essential lemon oil applied on rainbow trout fillets. They found that the coating limited lipid oxidation and displayed good antimicrobial activity after storage at 4°C for 15 days. Also, the use of water-soluble fraction of Farsi gum on the refrigerated shelf life of rainbow trout fillets was reported [159]. The texture, odor, color, and overall acceptability of coated samples were better than those observed in the control group. Also, the addition of cinnamon and thyme essential oils reduced the total viable, psychotropic, and acid bacteria values.

Chitosan is another natural polymer that is widely used as an ingredient to formulate edible films. The addition of chitosan to the batter to make enrobed fish sticks, and its effect on their functional properties and quality were evaluated. The presence of chitosan at 1% reduced the crispness, gumminess, and toughness of the product, and reduced lipid oxidation by ~65% [162].

1.4.4 ACTIVE PACKAGING IN MUSCLE FOODS

Continuous changes in consumer preferences for convenient, safe, healthy, and quality products drive innovations in food packaging systems. Edible films and coatings can be used as smart or active packaging, where the film/coating act as the carriers of food additives or components, such as an antioxidant, antimicrobial, O_2 scavengers, CO_2 emitters/absorbers, moisture regulator, flavor releasers, and absorbers agents [157]. Active packaging has been reported to delay or stop oxidation and enzymatic reactions, inhibit microbial growth, control the respiration rate, control weight loss, and retain the color and integrity of muscle foods [152].

Meat oxidation is one of the major causes of loss in meat products. Because of this, the incorporation of antioxidants into packaging films or coatings can enhance the stability of muscle foods. The presence of oxygen in the package promotes lipid oxidation, microbial growth, textural changes, discoloration, flavor development, toxic aldehydes, and nutritional losses as degradation of polyunsaturated fatty acids [157]. However, it is important to know the exact amount of antioxidant to use in edible films and coatings formulation. Higher amounts of antioxidant might induce the development of a heterogeneous structure, which triggers a decrease in the mechanical properties caused by the degree of protein-antioxidant interactions [163].

As mentioned above, muscle foods provide the optimal conditions for the growth of a wide range of microorganisms, thereby reducing the shelf life and increasing the risk of food-borne illness [164]. The most common microorganisms involved in muscle foods deterioration include microorganisms (such as *Salmonella* spp, *Mycobacterium* spp., *Staphylococcus aureus*, *Arcobacter butzleri*, *Listeria monocytogenes*, *Bacillus cereus*, *Clostridium botulinum*, *Clostridium perfringens*, *Yersinia enterocolitica*, *Escherichia coli* 0157:H7, *Aeromonas hydrophilla*, Enterohemorrhagic *E. coli* (EHEC) and *Campylobacter* spp.), spoilage bacteria (such as *Acinetobacter*, *Pseudomonas*, *Corynebacterium*, *Flavobacterium*, *Alcaligenes*, *Brochothrix thermosphacta*, *Moraxella*, *Klebsiella*, *Enterobacter*, *Proteus* spp., *Leuconostoc* spp. and *Lactobacillus* spp.) molds (commonly *Rhizopus*, *Aspergillus*, *Sporotichum*, *Monilia* and *Fusarium*), and yeasts (such as *Torulopsis* and *Candida* strains) [157]. Thus, the control of these microbial strains is important for prolonging the shelf life of muscle foods.

Several natural active compounds have been added directly to the polymers used for the packaging of meat products, resulting in increased shelf life. These compounds have included antimicrobial agents, plant extracts, essential oils, and lysozyme (Table 1.6).

Various antimicrobial agents and biopolymers can be used to elaborate edible films and coatings. Before applying active packaging to muscle foods, it is important to establish the proper antimicrobial agent. To do this, factors such as microbe target, possible interaction between polymer film, the antimicrobial compound and the food constituents, as well as their influence in sensory and visual characteristics of the packaged product, should be considered [157].

TABLE 1.6 Natural Components Used to Formulate Active Packaging in Meat, Fish, and Poultry Products

Food	Packaging components	Main results	Reference
Cold-smoked sardine	Gelatin + oregano or rosemary or chitosan	Increase in phenol content and antioxidant power of muscle decreasing the lipid oxidation and reducing microbial count.	[165]
Fresh chicken breast fillets	Alginate + thyme essential oil or propionic acid	Increase 33% the shelf life with the lowest dehydration, without significant changes in the sensorial analysis.	[166]
Chicken breast	Apple-pure-based films + carvacrol or cinnamaldehyde	Films with 3% antimicrobial showed reduction from 3.4 to 6.8 CFU/Log of *S. enterica* and *E. coli* O157:H7. Carvacrol induced greater reduction of *L. monocytogenes*	[167]
Fish fillets	Gelatin + tea polyphenols (TP)	Mesophylic and psychrotrophic count, yeast and molds were lower than those obtained with the control. The TP delayed the myosin and troponin degradation in myofibril of muscle foods.	[168]
Fish fillets	Nanoemulsions + rosemary, laurel, thyme and sage extracts	Increase in organoleptic quality, lowest bacterial count was obtained with rosemary and thyme extracts.	[169]
Nile tilapia fillets	Chitosan + pomegranate peel extract	A decrease in the entire microbial count during storage was observed. Edible coating increases the consumer's acceptability.	[170]
Mutton meat	Starch + clove and cinnamon extracts	Shelf life was increased 1 and 3 weeks at 10°C and 4°C, respectively. Edible coating reduced the growth of yeasts and molds; growth of *Pseudomonas*, *Lactobacilli* and *Biochothrix thermospacta* was delayed.	[142]
Sliced salmon	Catfish gelatin + chitosan and clove essential oil	Edible coating improved the antimicrobial activity, delaying the bacterial counts 2 log cycles at 2°C for 11 days.	[163]

1.4.5 METHODS OF COATING APPLICATION ON MUSCLE FOODS

Edible coatings can be applied to meat and fish products by multiple methods, including dipping, foaming, spraying, casting, brushing or spraying, and individual wrapping or rolling [171, 172]. Other methods including as fluidization, UV polymerization, enrobing and extrusion have been reported to manufacture whey protein-based films and coatings [173]. These methods have several advantages and disadvantages, and their selection depends on the desired product, coating thickness, solution rheology, and drying technique [174].

1.5 EXTENDING SHELF LIFE OF INTERMEDIATE MOISTURE AND THERMALLY PROCESSED FOODS WITH EDIBLE FILMS AND COATINGS

1.5.1 APPLICATION ON INTERMEDIATE MOISTURE FOODS

Intermediate moisture foods are defined as some products that are stabilized by their reduced water activity [175]. These products include derivate dried meat [176], sausage [77], ham [177], tuna fish [178], Port Salut cheese [179], Mozzarella cheese [171], Saloio cheese [180], nuts [181], almonds [182], and bakery products [183]. These products are generally highly perishable, which is driving interest in the use of biodegradable films and coatings. The inclusion of additives improves the properties of the edible films and coatings by the donating hydrogen atoms, trapping free radicals, interrupting chain oxidation reactions, and chelating metal ions [184]. Table 1.7 summarizes the most recent application and effect of an edible films and coatings on intermediate moisture foods.

1.5.2 APPLICATION ON THERMALLY PROCESSED FOODS

Thermal processing is a common method used for industrial food production. Thermally processed foods are ham [177], salami and frankfurters [190]. Diverse additives have been used to improve the shelf life of these

TABLE 1.7 Edible Films and Coating Applied on Intermediate Moisture Foods

Product	Components	Bioactive compounds	Variables	Main results	Reference
Cashew nut	Starch, CTG,	–	WVP	The presence of MMT increased coating protection against moisture loss, suggesting that the nanocomposite coating is more effective than the starch-CTG coating to extend the stability of cashew nuts.	[181]
Nut	CS	GTE	PV, TAV, M, SE, St.S	The GTE into CS reduce oxidation activity and fungal growth	[185]
Ham	DSM	LOPS	IC, HA	The LPOS-DSM film coating is promising for controlling salmonellosis associated with consumption of ham.	[186]
Sausage	CS	CO	M, CA, pH LE, PV TAV, SE	Experiments demonstrated that the combination of CS and clove oil inhibited microbial growth, retarded lipid oxidation, and extended the shelf life of cooked pork sausages in refrigerated storage. However, there were some initially negative impacts on odor and taste qualities, at the start of storage	[187]
Mini panettones	PS	SUC	SE, W, M, T, IS	The incorporation of food-graded antimicrobial compounds in the packaging films of potato starch coatings in concentrations lower than those normally used for mini panettones increased up to 130% their shelf life.	[188]
Fish Sausage	CS, Gelatin, GLY	Lactic acid	ANA; WC; WHC; SE	The coating increased the delay phase of total viable microorganisms and enterobacteria at 15 and 10 days, respectively, while the film drastically inhibited the growth of these groups	[77]
Sausage	WPC	EO	CA, TPA, pH, MC	Extended your life approximately 15–20 days	[189]

CTG: Cashew tree gum, WVP: Water vapor permeability, MMT: Montmorillonite, CS: Chitosan, GTE: Green tea extract, PV: Peroxide value, TAV: Thiobarbituric acid value, M: Microbiology, SE: Sensory evaluation, St.S: Storange stability, DSM: Defatted soy bean meal, LOPS: Lactoperoxidase system, IC: Inhibitory concentration, HA: Hipotiocianita analysis, CO: Clove oil, LE: Lipid extraction, PV: Peroxide value, CA: Color analysis, PS: Potato starch, IS: Inverted sugar, SUC: Sucrose, WA: Water activity, T: Texture, ANA: Antimicrobial activity, WC: Water content, WHC: Water holding capacity, GLY: Glycerol, EO: *Origanum virens* essential, WPC: Whey protein active, MC: Moisture contents.

products. These foods are subjected to a thermal treatment to cooking and eliminate pathogens; however, during peeling, slicing or packaging, they are susceptible to post-process contamination. Because of this, the FDA (Food and Drug Administration) and USDA (United States Department of Agriculture) have classified non-reheated frankfurters as a product of high risk [177]. Also, the intensity of the thermal treatments is a matter of consumer choice, creating a risk of inadequate eliminations of pathogenic microorganisms. For this reason, the use of active edible films and coatings with antimicrobial agents is a promising method for controlling spoilage and photogenic microorganisms [177]. Table 1.8 shows the most recent use and application of an edible films and coatings on thermally processed foods.

1.5.3 METHODS OF COATING APPLICATION ON INTERMEDIATE MOISTURE AND THERMALLY PROCESSED FOODS

Edible coatings can be applied on or even within foods by various methods. For example, dipping is the most common lab-scale method due to its simplicity, low cost and homogeneity. However, this method generates a residual coating material that contributes to microbial growth [171]. However, this method has some disadvantages, for example, leads to a high quantity of residual coating materials, and often results in microorganism growth in the dipping tank. Because of this, it is important to know the viscosity and density of coating solutions, and surface tension of coated products [174]. Spray coating is the most commonly used technique for food application [193]. Spray application increases the surface area of the liquid through the formation of droplets and distributes them over the food surface area by means of a set of nozzles. This technique offers uniform coating thickness, temperature control and facilitates the automation of continuous production [174]. Another technique is Fluidized-bed, which can be used to apply a very thin layer. This technique involves spraying through a set of nozzles onto the surface of fluidized powders to form a shell-type structure and the conventional top-spray method has a greater possibility of success in the food industry compared to other methods [171]. Finally, the planning process consists of placing

Edible Active Coatings for Foods as a Key Factor for Shelf-Life Prolongation 27

TABLE 1.8 Edible Films and Coating Applied on Thermally Processed Foods

Product	Components	Bioactive compounds	Variables	Main results	Reference
Ham slices	Na-alginate, GLY.	"tsipouro" (41% v/v ethanol), "rak" or "tsikoudia" (39.6% v/v ethanol) and "ouzo" (40% v/v ethanol)	MA	Maintain levels of L. monocytogenes at least 5.0 log CFU/cm2 lower (p < 0.05) than controls after 81 and 52 days at 4 and 10°C	[177]
Ham slices	Soybean meal, xanthan, GLY	Oil extract from soybean lactoperxidase (81 U/mg)	MA	The film containing OSCN− at 0.66 mg/g could provide 42.9 h of protection against 2.8 log CFU/g of Salmonella Typhimurium.	[186]
Frankfurters	Chitosan,	Trans-cinnamaldehyde (0.1, 0.2%), carvacrol (0.1, 0.2%), thymol (0.1, 0.2%) and eugenol (0.3, 0.4%)	Lipid oxidation, pH and microbiological analysis.	All phytochemical coatings reduced LM counts by >2.5 log CFU/frankfurter compared to controls (P < 0.05) and reduced the lipid oxidation. No change in pH was observed.	[190]
Dry-cured hams	Lard, mineral oil, GLY, PGXG, carragenan	PS	SE	Cubes coated with xanthan gum + 20% propylene glycol and carrageenan/propylene glycol alginate + 10% propylene glycol were effective at controlling mite infestations under laboratory conditions	[191]
Meat sousage	CGS	CGS	Physicochemical properties, water holding capacity, textural properties, color and sensorial properties.	Water holding capacity, hardness, adhesivity and chewiness of the formulated sausage samples and contributed to the final product lightness. Gelatin addition had no significant effect on sausages taste using trained panel. Further, sausage slice ability, texture and global acceptability were markedly improved.	[192]

MA: Microbiological analysis, GLY: Glycerol, PS: Potassium sorbate, PGXG: Propylene glycol xanthan gum, CGS: Cuttlefish skin gelatin, SE: Sensory evaluation.

the product into a large, rotating bowl, referred to as the pan. Forced air, either ambient or at elevated temperature is then applied to dry the coating [194]. Each technique has advantages and disadvantages. The selection of application method depends directly upon the desired product, thickness, solution rheology and drying technology, thereby facilitating factory-scale implementation.

1.6 CONCLUSIONS

Edible films and coatings have enormous potential to extend the shelf life of fresh and processed foods by maintaining their quality, safety, and functionality during storage. Moreover, these materials can perform as active packaging incorporating antioxidants and antimicrobial compounds to reduce the oxidation and microbial contamination, respectively. Another advantage is that edible films and coatings are considered biodegradable packages and, therefore, important eco-friendly alternatives to synthetic polymeric packing. However, their mechanical properties and predominantly hydrophilic character limit the commercial application of an edible films and coatings in food packaging. Therefore, the greatest challenge in food packaging research is to develop and improve the mechanical and water barrier properties of an edible films and coatings so as to make them economically and practically viable.

KEYWORDS

- active coatings
- edible coatings
- food quality
- prolongation
- shelf life

REFERENCES

1. Irmak, S., & Erbatur, O., (2008). Additives for environmentally compatible active food packaging. In: Chiellini, E., (ed.), *Environmentally Compatible Food Packaging* (pp. 263–293), CRC Press LLC, Parkway, NW.

2. Pavlath, A. E., & Orts, W., (2009). Edible films and coatings: Why, what, and how? In: Embuscado, M. E., & Huber, K. C., (eds.), *Edible Films and Coatings for Food Applications* (pp. 1–23), Springer: New York, USA.
3. Hardenburg, R. E., (1967). *Wax and Related Coatings for Horticultural Products, a Bibliography*, Research Service Bulletin 51-15, United States Department of Agriculture, Washington, DC, pp 1–15.
4. Biquet, B., & Guilbert, S., (1986). Relative diffusivities of water in model intermediate moisture foods. *Lebensm. Wiss. Technol.*, *19*(3), 208–214.
5. Havard, C., & Harmony, M. X., (1869). Improved process of preserving meat, fowls, fish. *Patent 90, 944*.
6. Umaraw, P., & Verma, A. K., (2017). Comprehensive review on application of an edible film on meat and meat products: An eco-friendly approach. *Crit. Rev. Food Sci. Nutr.*, *57*(6), 1270–1279.
7. Contreras-Medellin, R., & Labuza, T. P., (1981). Prediction of moisture protection requirements for foods. *Cereal Food World*, *27*(7), 335–340.
8. Earle, R. D., & Snyder, C. E., (1966). Method of preparing frozen seafood. *Patent 3, 255, 021*.
9. Bauer, C. D., Neuser, G. L., & Pinkalla, H. A., (1968). Method for preparing a coated meat product. *Patent 3, 406, 081*.
10. Galus, S., & Kadzińska, J., (2015). Food applications of emulsion-based edible films and coatings. *Trends Food Sci. Technol.*, *45*(2), 273–283.
11. Ramos, Ó. L., Fernandes, J. C., Silva, S. I., Pintado, M. E., & Malcata, F. X., (2012). Edible films and coatings from whey proteins: A review on formulation, and on mechanical and bioactive properties. *Crit. Rev. Food Sci. Nutr.*, *52*(6), 533–552.
12. Sothornvit, R., (2015). Edible films and coatings for packaging applications. In: Alavi, S., Thomas, S., Sandeep, K. P., Kalarikkal, N., Varghese, J., & Yaragalla, S., (eds.), *Polymers for Packaging Applications* (pp. 173–196.). CRC Press LLC, Parkway, NW.: Ontario, Canada.
13. Krochta, J. M., (2002). Proteins as raw materials for films and coatings: Definitions, current status, and opportunities. In: Gennadios, A., (ed.), *Protein-Based Films and Coatings* (pp. 1–41). CRC Press: Boca Raton, FL, USA.
14. Khwaldia, K., Perez, C., Banon, S., Desobry, S., & Hardy, J., (2004). Milk proteins for edible films and coatings. *Crit. Rev. Food Sci. Nutr.*, *44*(4), 239–251.
15. Cazón, P., Velazquez, G., Ramírez, J. A., & Vázquez, M., (2017). Polysaccharide-based films and coatings for food packaging: A review. *Food Hydrocoll.*, *68*, 136–148.
16. Debeaufort, F., Quezada-Gallo, J. A., & Voilley, A., (1998). Edible films and coatings: Tomorrow's packagings: A review. *Crit. Rev. Food Sci. Nutr.*, *38*(4), 299–313.
17. Guilbert, S., Gontard, N., & Cuq, B., (1995). Technology and applications of an edible protective films. *Packag. Technol. Sci.*, *8*(6), 339–346.
18. Hernandez-Izquierdo, V. M., & Krochta, J. M., (2008). Thermoplastic processing of proteins for film formation-A review. *J. Food Sci.*, *73*(2), 30–39.
19. Valenzuela, C., Abugoch, L., & Tapia, C., (2013). Quinoa protein–chitosan–sunflower oil edible film: Mechanical, barrier and structural properties. *LWT-Food Sci. Technol.*, *50*(2), 531–537.

20. Han, J. H., (2014). Edible films and coatings: A review. In: Han, J. H., (ed.), *Innovations in Food Packaging* (2nd edn., pp. 213–255). Academic Press: San Diego, CA, USA.
21. Elsabee, M. Z., & Abdou, E. S., (2013). Chitosan-based edible films and coatings: A review. *Mater Sci. Eng. C.*, *33*(4), 1819–1841.
22. Espitia, P. J. P., Du, W. X., Avena-Bustillos, R. D. J., Soares, N. D. F. F., & McHugh, T. H., (2014). Edible films from pectin: Physical-mechanical and antimicrobial properties – A review. *Food Hydrocoll.*, *35*, 287–296.
23. Jiménez, A., Fabra, M. J., Talens, P., & Chiralt, A., (2012). Edible and biodegradable starch films: A review. *Food Bioproc. Tech.*, *5*(6), 2058–2076.
24. Xu, Q., Chen, C., Rosswurm, K., Yao, T., & Janaswamy, S., (2016). A facile route to prepare cellulose-based films. *Carbohydr. Polym.*, *149*, 274–281.
25. Pérez-Gago, M. B., & Rhim, J. W., (2014). Edible coating and film materials: Lipid bilayers and lipid emulsions. In: Han, J. H., (ed.), *Innovations in Food Packaging* (2nd edn., pp. 325–345). Academic Press, London: San Diego, CA, USA.
26. Kurek, M., Ščetar, M., & Galić, K., (2017). Edible coatings minimize fat uptake in deep fat fried products: A review. *Food Hydrocoll*, *71*, 225–235.
27. Guilbert, S., Cuq, B., & Gontard, N., (1997). Recent innovations in edible and/or biodegradable packaging materials. *Food Add. Contam.*, *14*(6–7), 741–751.
28. Vieira, M. G. A., Da Silva, M. A., Dos Santos, L. O., & Beppu, M. M., (2011). Natural-based plasticizers and biopolymer films: A review. *Eur. Polym. J.*, *47*(3), 254–263.
29. Kang, H. J., Kim, S. J., You, Y. S., Lacroix, M., & Han, J., (2013). Inhibitory effect of soy protein coating formulations on walnut (*Juglans regia* L.) kernels against lipid oxidation. *LWT-Food Sci. Technol.*, *51*(1), 393–396.
30. Lee, H. B., Noh, B. S., & Min, S. C., (2012). Listeria monocytogenes inhibition by defatted mustard meal-based edible films. *Int. J. Food Microbiol.*, *153*(1), 99–105.
31. Bourtoom, T., (2008). Edible films and coatings: Characteristics and properties. *Food Res. Int.*, *15*(3), 237–248.
32. Kester, J. J., & Fennema, O. R., (1986). Edible films and coatings: A review. *Food Technol.40*, 47–59.
33. Manrich, A., Moreira, F. K. V., Otoni, C. G., Lorevice, M. V., Martins, M. A., & Mattoso, L. H. C., (2017). Hydrophobic edible films made up of tomato cutin and pectin. *Carbohydr. Polym.*, *164*, 83–91.
34. Mahalik, N. P., & Nambiar, A. N., (2010). Trends in food packaging and manufacturing systems and technology. *Trends Food Sci. Technol.*, *21*(3), 117–128.
35. Sutherland, W. J., Clout, M., Côté, I. M., Daszak, P., Depledge, M. H., Fellman, L., et al., (2010). A horizon scan of global conservation issues for 2010. *Trends Ecol. Evol.*, *25*(1), 1–7.
36. Sessini, V., Arrieta, M. P., Kenny, J. M., & Peponi, L., (2016). Processing of an edible films based on nanoreinforced gelatinized starch. *Polym. Degrad. Stab.*, *132*, 157–168.
37. Guimarães, I. C., Dos Reis, K. C., Menezes, E. G. T., Rodrigues, A. C., Da Silva, T. F., De Oliveira, I. R. N., et al., (2016). Cellulose microfibrillated suspension of carrots obtained by mechanical defibrillation and their application in edible starch films. *Ind. Crops Prod.*, *89*, 285–294.

38. Sahraee, S., Milani, J. M., Ghanbarzadeh, B., & Hamishehkar, H., (2017). Physicochemical and antifungal properties of bio-nanocomposite film based on gelatin-chitin nanoparticles. *Int. J. Biol. Macromol.*, *97*, 373–381.
39. Murmu, S. B., & Mishra, H. N., (2017). Optimization of the arabic gum based edible coating formulations with sodium caseinate and tulsi extract for guava. *LWT-Food Sci. Technol.*, *80*, 271–279.
40. Tavassoli-Kafrani, E., Shekarchizadeh, H., & Masoudpour-Behabadi, M., (2016). Development of an edible films and coatings from alginates and carrageenans. *Carbohydr. Polym.*, *137*, 360–374.
41. Fakhouri, F. M., Martelli, S. M., Caon, T., Velasco, J. I., & Mei, L. H. I., (2015). Edible films and coatings based on starch/gelatin: Film properties and effect of coatings on quality of refrigerated red crimson grapes. *Postharvest Biol. Technol.*, *109*, 57–64.
42. Liu, X., Tang, C., Han, W., Xuan, H., Ren, J., Zhang, J., & Ge, L., (2017). Characterization and preservation effect of polyelectrolyte multilayer coating fabricated by carboxymethyl-cellulose and chitosan. *Colloids Surf. A Physicochem. Eng. Asp.* *529*, 1016–1023.
43. Cardoso, G. P., Dutra, M. P., Fontes, P. R., Ramos, A. D. L. S., Gomide, L. A. D. M., & Ramos, E. M., (2016). Selection of a chitosan gelatin-based edible coating for color preservation of beef in retail display. *Meat Sci.*, *114*, 85–94.
44. Limchoowong, N., Sricharoen, P., Techawongstien, S., & Chanthai, S., (2016). An iodine supplementation of tomato fruits coated with an edible film of the iodide-doped chitosan. *Food Chem.*, *200*, 223–229.
45. Treviño-Garza, M. Z., García, S., Heredia, N., Alanís-Guzmán, M. G., & Arévalo-Niño, K., (2017). Layer-by-layer edible coatings based on mucilages, pullulan and chitosan and its effect on quality and preservation of fresh-cut pineapple (*Ananas comosus*). *Postharvest Biol. Technol.*, *128*, 63 75.
46. Allegra, A., Sortino, G., Inglese, P., Settanni, L., Todaro, A., & Gallotta, A., (2017). The effectiveness of *Opuntia ficus-indica* mucilage edible coating on post-harvest maintenance of 'Dottato' fig (*Ficus carica* L.) fruit. *Food Packaging and Shelf Life*, *12*, 135–141.
47. Allegra, A., Inglese, P., Sortino, G., Settanni, L., Todaro, A., & Liguori, G., (2016). The influence of *Opuntia ficus-indica* mucilage edible coating on the quality of 'Hayward' kiwifruit slices. *Postharvest Biol. Technol.*, *120*, 45–51.
48. Sharma, L., & Singh, C., (2016). Sesame protein based edible films: Development and characterization. *Food Hydrocoll.*, *61*, 139–147.
49. Wihodo, M., & Moraru, C. I., (2015). Effect of pulsed light treatment on the functional properties of casein films. *LWT-Food Sci. Technol.*, *64*(2), 837–844.
50. Musso, Y. S., Salgado, P. R., & Mauri, A. N., (2017). Smart edible films based on gelatin and curcumin. *Food Hydrocoll.*, *66*, 8–15.
51. López, D., Márquez, A., Gutiérrez-Cutiño, M., Venegas-Yazigi, D., Bustos, R., & Matiacevich, S., (2017). Edible film with antioxidant capacity based on salmon gelatin and boldine. *LWT-Food Sci. Technol.*, *77*, 160–169.
52. Galus, S., & Kadzińska, J., (2016). Whey protein edible films modified with almond and walnut oils. *Food Hydrocoll.*, *52*, 78–86.

53. Akcan, T., Estévez, M., & Serdaroğlu, M., (2017). Antioxidant protection of cooked meatballs during frozen storage by whey protein edible films with phytochemicals from *Laurus nobilis L.* and *Salvia officinalis*. *LWT-Food Sci. Technol.*, *77*, 323–331.
54. Escamilla-García, M., Calderón-Domínguez, G., Chanona-Pérez, J. J., Farrera-Rebollo, R. R., Andraca-Adame, J. A., Arzate-Vázquez, I., Mendez-Mendez, J. V., & Moreno-Ruiz, L. A., (2013). Physical and structural characterisation of zein and chitosan edible films using nanotechnology tools. *Innov. Food Sci. Emerg. Technol.*, *61*, 196–203.
55. Moradi, M., Tajik, H., Razavi Rohani, S. M., & Mahmoudian, A., (2016). Antioxidant and antimicrobial effects of zein edible film impregnated with *Zataria multiflora* Boiss. essential oil and monolaurin. *LWT-Food Sci. Technol.*, *72*, 37–43.
56. Tulamandi, S., Rangarajan, V., Rizvi, S. S. H., Singhal, R. S., Chattopadhyay, S. K., & Saha, N. C., (2016). A biodegradable and edible packaging film based on papaya puree, gelatin, and defatted soy protein. *Food Packaging and Shelf Life*, *10*, 60–71.
57. Alves, M. M., Gonçalves, M. P., & Rocha, C. M. R., (2017). Effect of ferulic acid on the performance of soy protein isolate-based edible coatings applied to fresh-cut apples. *LWT-Food Sci. Technol.*, *80*, 409–415.
58. Rocca-Smith, J. R., Marcuzzo, E., Karbowiak, T., Centa, J., Giacometti, M., Scapin, F., Venir, E., Sensidoni, A., & Debeaufort, F., (2016). Effect of lipid incorporation on functional properties of wheat gluten based edible films. *J. Cereal Sci.*, *69*, 275–282.
59. Ansorena, M. R., Zubeldía, F., & Marcovich, N. E., (2016). Active wheat gluten films obtained by thermoplastic processing. *LWT-Food Sci. Technol.*, *69*, 47–54.
60. Reddy, N., Jiang, Q., & Yang, Y., (2012). Preparation and properties of peanut protein films crosslinked with citric acid. *Ind. Crops Prod.*, *39*, 26–30.
61. Sun, Q., Sun, C., & Xiong, L., (2013). Mechanical, barrier and morphological properties of pea starch and peanut protein isolate blend films. *Carbohydr. Polym.*, *98*(1), 630–637.
62. Rompothi, O., Pradipasena, P., Tananuwong, K., Somwangthanaroj, A., & Janjarasskul, T., (2017). Development of non-water soluble, ductile mung bean starch based edible film with oxygen barrier and heat sealability. *Carbohydr. Polym.*, *157*, 748–756.
63. Bonilla, J., & Sobral, P. J. A., (2016). Investigation of the physicochemical, antimicrobial and antioxidant properties of gelatin-chitosan edible film mixed with plant ethanolic extracts. *Food Biosci.*, *16*, 17–25.
64. Ramos, Ó. L., Reinas, I., Silva, S. I., Fernandes, J. C., Cerqueira, M. A., Pereira, R. N., Vicente, A. A., Poças, M. F., Pintado, M. E., & Malcata, F. X., (2013). Effect of whey protein purity and glycerol content upon physical properties of an edible films manufactured therefrom. *Food Hydrocoll.*, *30*(1), 110–122.
65. Vonasek, E. Le, P., & Nitin, N., (2014). Encapsulation of bacteriophages in whey protein films for extended storage and release. *Food Hydrocoll.*, *37*, 7–13.
66. Fadini, A. L., Rocha, F. S., Alvim, I. D., Sadahira, M. S., Queiroz, M. B., Alves, R. M. V., & Silva, L. B., (2013). Mechanical properties and water vapor permeability of hydrolysed collagen–cocoa butter edible films plasticised with sucrose. *Food Hydrocoll.*, *30*(2), 625–631.

67. Wang, K., Wang, W., Ye, R., Liu, A., Xiao, J., Liu, Y., & Zhao, Y., (2017). Mechanical properties and solubility in water of corn starch-collagen composite films: Effect of starch type and concentrations. *Food Chem.*, *216*, 209–216.
68. Güçbilmez, Ç. M., Yemenicioğlu, A., & Arslanoğlu, A., (2007). Antimicrobial and antioxidant activity of an edible zein films incorporated with lysozyme, albumin proteins and disodium EDTA. *Food Res. Int.*, *40* (1), 80–91.
69. Gómez-Estaca, J., Montero, P., & Gómez-Guillén, M. C., (2014). Shrimp (*Litopenaeus vannamei*) muscle proteins as source to develop edible films. *Food Hydrocoll.*, *41*, 86–94.
70. Leerahawong, A., Arii, R., Tanaka, M., & Osako, K., (2011). Edible film from squid (*Todarodes pacificus*) mantle muscle. *Food Chem.*, *124*(1), 177–182.
71. Lee, J. H., Won, M., & Song, K. B., (2015). Physical properties and antimicrobial activities of porcine meat and bone meal protein films containing coriander oil. *LWT-Food Sci. Technol.*, *63*(1), 700–705.
72. Voon, H. C., Bhat, R., Easa, A. M., Liong, M. T., & Karim, A. A., (2012). Effect of addition of halloysite nanoclay and SiO2 nanoparticles on barrier and mechanical properties of bovine gelatin films. *Food Bioproc. Tech.*, *5*(5), 1766–1774.
73. Boran, G., & Regenstein, J. M., (2010). Fish gelatin. *Adv. Food Nutr. Res.*, *60*, 119–143.
74. Lima, Á. M., Cerqueira, M. A., Souza, B. W. S., Santos, E. C. M., Teixeira, J. A., Moreira, R. A., & Vicente, A. A., (2010). New edible coatings composed of galactomannans and collagen blends to improve the postharvest quality of fruits – Influence on fruits gas transfer rate. *J. Food Eng.*, *97*(1), 101–109.
75. Mannucci, A., Serra, A., Remorini, D., Castagna, A., Mele, M., Scartazza, A., et al., (2017). Aroma profile of Fuji apples treated with gelatin edible coating during their storage. *LWT-Food Sci. Technol.*, *85*, 28–36.
76. Boyacı, D., Korel, F., & Yemenicioğlu, A., (2016). Development of activate-at-home-type edible antimicrobial films: An example pH-triggering mechanism formed for smoked salmon slices using lysozyme in whey protein films. *Food Hydrocoll.*, *60*, 170–178.
77. Alemán, A., González, F., Arancibia, M. Y., López-Caballero, M. E., Montero, P., & Gómez-Guillén, M. C., (2016). Comparative study between film and coating packaging based on shrimp concentrate obtained from marine industrial waste for fish sausage preservation. *Food Control*, *70*, 325–332.
78. Song, N. B., Lee, J. H., Al Mijan, M., & Song, K. B., (2014). Development of a chicken feather protein film containing clove oil and its application in smoked salmon packaging. *LWT-Food Sci. Technol.*, *57*(2), 453–460.
79. Marcuzzo, E., Peressini, D., Debeaufort, F., & Sensidoni, A., (2010). Effect of ultrasound treatment on properties of gluten-based film. *Innov. Food Sci. Emerg. Technol.*, *11*(3), 451–457.
80. Chen, F., Monnier, X., Gällstedt, M., Gedde, U. W., & Hedenqvist, M. S., (2014). Wheat gluten/chitosan blends: A new biobased material. *Eur. Polym. J.*, *60*, 186–197.
81. Gaona-Sánchez, V. A., Calderón-Domínguez, G., Morales-Sánchez, E., Chanona-Pérez, J. J., Velázquez-de la Cruz, G., Méndez-Méndez, J. V., et al., (2015). Prepara-

tion and characterisation of zein films obtained by electrospraying. *Food Hydrocoll.*, *49*, 1–10.

82. Aydemir, L. Y., Gökbulut, A. A., Baran, Y., & Yemenicioğlu, A., (2014). Bioactive, functional and edible film-forming properties of isolated hazelnut (Corylus avellana L.) meal proteins. *Food Hydrocoll.*, *36*, 130–142.

83. Arabestani, A., Kadivar, M., Shahedi, M., Goli, S. A. H., & Porta, R., (2016). Characterization and antioxidant activity of bitter vetch protein-based films containing pomegranate juice. *LWT-Food Sci. Technol.*, *74*, 77–83.

84. Porta, R., Di Pierro, P., Rossi-Marquez, G., Mariniello, L., Kadivar, M., & Arabestani, A., (2015). Microstructure and properties of bitter vetch (*Vicia ervilia*) protein films reinforced by microbial transglutaminase. *Food Hydrocoll.*, *50*, 102–107.

85. Liu, P., Xu, H., Zhao, Y., & Yang, Y., (2017). Rheological properties of soy protein isolate solution for fibers and films. *Food Hydrocoll.*, *64*, 149–156.

86. Abugoch, L. E., Tapia, C., Villamán, M. C., Yazdani-Pedram, M., & Díaz-Dosque, M., (2011). Characterization of quinoa protein–chitosan blend edible films. *Food Hydrocoll.*, *25*(5), 879–886.

87. Alonso, A., Beunza, J. J., Bes-Rastrollo, M., Pajares, R. M., & Martínez-González, M. Á., (2006). Vegetable protein and fiber from cereal are inversely associated with the risk of hypertension in a spanish cohort. *Arch. Med. Res.*, *37*(6), 778–786.

88. Azeredo, H. M. C., & Waldron, K. W., (2016). Cross-linking in polysaccharide and protein films and coatings for food contact – A review. *Trends Food Sci. Technol.*, *52*, 109–122.

89. Brandenburg, A. H., Weller, C. L., & Testin, R. F., (1993). Edible films and coatings from soy protein. *J. Food Sci.*, *58*(5), 1086–1089.

90. Pena-Serna, C., Penna, A. L. B., & Lopes Filho, J. F., (2016). Zein-based blend coatings: Impact on the quality of a model cheese of short ripening period. *J. Food Eng.*, *171*, 208–213.

91. Valenzuela, C., Tapia, C., López, L., Bunger, A., Escalona, V., & Abugoch, L., (2015). Effect of an edible quinoa protein-chitosan-based films on refrigerated strawberry (*Fragaria×ananassa*) quality. *Electron. J. Biotechnol.*, *18*(6), 406–411.

92. Chinma, C. E., Ariahu, C. C., & Abu, J. O., (2014). Shelf life extension of toasted groundnuts through the application of cassava starch and soy protein-based edible coating. *Nigeria Food, J., 32*(1), 133–138.

93. Aguirre-Joya, J., lvarez, B., Ventura, J., García-Galindo, J., De León-Zapata, M., Rojas, R., etal., (2016). Edible coatings and films from lipids, waxes, and resins. In: Ribeiro, C. M. Â. P., Correia, P. R. N., Da Silva, R. Ó. L., Couto, T. J. A., & Augusto, V. A., (eds.), *Edible Food Packaging* (pp. 121–152). CRC Press: Boca Raton, FL, USA.

94. Kowalczyk, D., & Baraniak, B., (2014). Effect of candelilla wax on functional properties of biopolymer emulsion films – A comparative study. *Food Hydrocoll.*, *41*, 195–209.

95. Oregel-Zamudio, E., Angoa-Pérez, M. V., Oyoque-Salcedo, G., Aguilar-González, C. N., & Mena-Violante, H. G., (2017). Effect of candelilla wax edible coatings combined with biocontrol bacteria on strawberry quality during the shelf life. *Sci. Hortic.*, *214*, 273–279.

96. Saucedo-Pompa, S., Rojas-Molina, R., Aguilera-Carbó, A. F., Saenz-Galindo, A., Garza, H. D. L., Jasso-Cantú, D., et al., (2009). Edible film based on candelilla wax to improve the shelf life and quality of avocado. *Food Res. Int., 42*(4), 511–515.
97. Chiumarelli, M., & Hubinger, M. D., (2014). Evaluation of an edible films and coatings formulated with cassava starch, glycerol, carnauba wax and stearic acid. *Food Hydrocoll., 38*, 20–27.
98. Rodrigues, D. C., Caceres, C. A., Ribeiro, H. L., De Abreu, R. F. A., Cunha, A. P., & Azeredo, H. M. C., (2014). Influence of cassava starch and carnauba wax on physical properties of cashew tree gum-based films. *Food Hydrocoll., 38*, 147–151.
99. Soradech, S., Nunthanid, J., Limmatvapirat, S., & Luangtana-anan, M., (2017). Utilization of shellac and gelatin composite film for coating to extend the shelf life of banana. *Food Control, 73*, 1310–1317.
100. Khorram, F., Ramezanian, A., & Hosseini, S. M. H., (2017). Shellac, gelatin and persian gum as alternative coating for orange fruit. *Sci. Hortic., 225*, 22–28.
101. Haq, M. A., Hasnain, A., Jafri, F. A., Akbar, M. F., & Khan, A., (2016). Characterization of an edible gum cordia film: Effects of beeswax. *LWT-Food Sci. Technol., 68*, 674–680.
102. Saurabh, C. K., Gupta, S., Variyar, P. S., & Sharma, A., (2016). Effect of addition of nanoclay, beeswax, tween-80 and glycerol on physicochemical properties of guar gum films. *Ind. Crops Prod., 89*, 109–118.
103. Jafari, S. M., Khanzadi, M., Mirzaei, H., Dehnad, D., Chegini, F. K., & Maghsoudlou, Y., (2015). Hydrophobicity, thermal and micro-structural properties of whey protein concentrate–pullulan–beeswax films. *Innov. Food Sci. Emerg. Technol., 80*, 506–511.
104. Shih, F. F., Daigle, K. W., & Champagne, E. T., (2011). Effect of rice wax on water vapor permeability and sorption properties of an edible pullulan films. *Food Chem., 127*(1), 118–121.
105. Hashemi, S. M. B., & Mousavi Khaneghah, A., (2017). Characterization of novel basil-seed gum active edible films and coatings containing oregano essential oil. *Prog. Org. Coat., 110*, 35–41.
106. Rodrigues, D. C., Cunha, A. P., Brito, E. S., Azeredo, H. M. C., & Gallão, M. I., (2016). Mesquite seed gum and palm fruit oil emulsion edible films: Influence of oil content and sonication. *Food Hydrocoll., 56*, 227–235.
107. Sánchez Aldana, D., Andrade-Ochoa, S., Aguilar, C. N., Contreras-Esquivel, J. C., & Nevárez-Moorillón, G. V., (2015). Antibacterial activity of pectic-based edible films incorporated with Mexican lime essential oil. *Food Control, 50*, 907–912.
108. Riquelme, N., Herrera, M. L., & Matiacevich, S., (2017). Active films based on alginate containing lemongrass essential oil encapsulated: Effect of process and storage conditions. *Food Bioprod. Process., 104*, 94–103.
109. Ma, W., Tang, C. H., Yin, S. W., Yang, X. Q., Wang, Q., Liu, F., et al., (2012). Characterization of gelatin-based edible films incorporated with olive oil. *Food Res. Int., 49*(1), 572–579.
110. Ali, A., Maqbool, M., Alderson, P. G., & Zahid, N., (2013). Effect of gum arabic as an edible coating on antioxidant capacity of tomato (*Solanum lycopersicumL.*) fruit during storage. *Postharvest Biol. Technol., 76*, 119–124.

111. Mostafavi, F. S., Kadkhodaee, R., Emadzadeh, B., & Koocheki, A., (2016). Preparation and characterization of tragacanth–locust bean gum edible blend films. *Carbohydr. Polym.*, *139*, 20–27.
112. Zhang, P., Zhao, Y., & Shi, Q., (2016). Characterization of a novel edible film based on gum ghatti: Effect of plasticizer type and concentration. *Carbohydr. Polym.*, *153*, 345–355.
113. Singh, S., Khemariya, P., Rai, A., Rai, A. C., Koley, T. K., & Singh, B., (2016). Carnauba wax-based edible coating enhances shelf life and retain quality of eggplant (*Solanum melongena*) fruits. *LWT-Food Sci. Technol.*, *74*, 420–426.
114. Choi, W. S., Singh, S., & Lee, Y. S., (2016). Characterization of an edible film containing essential oils in hydroxypropyl methylcellulose and its effect on quality attributes of 'Formosa' plum (*Prunus salicina* L.). *LWT-Food Sci. Technol.*, *70*, 213–222.
115. Artiga-Artigas, M., Acevedo-Fani, A., & Martín-Belloso, O., (2017). Improving the shelf life of low-fat cut cheese using nanoemulsion-based edible coatings containing oregano essential oil and mandarin fiber. *Food Control*, *76*, 1–12.
116. Khalifa, I., Barakat, H., El-Mansy, H. A., & Soliman, S. A., (2016). Improving the shelf life stability of apple and strawberry fruits applying chitosan-incorporated olive oil processing residues coating. *Food Packaging and Shelf Life*, *9*, 10–19.
117. Tafolla-Arellano, J. C., González-León, A., Tiznado-Hernández, M. E., García, L. Z., & Báez-Sañudo, R., (2013). Composición, fisiología y biosíntesis de la cutícula en Plantas. *Rev. Fitotec. Mex.*, *36*(1), 3–12.
118. Lin, D., & Zhao, Y., (2007). Innovations in the development and application of an edible coatings for fresh and minimally processed fruits and vegetables. *Compr. Rev. Food Sci. Food Saf.*, *6*(3), 60–75.
119. Guilbert, S., Gontard, N., & Gorris, L. G. M., (1996). Prolongation of the shelf life of perishable food products using biodegradable films and coatings. *LWT-Food Sci. Technol.*, *29*(1), 10–17.
120. Artes, F., & Allende, A., (2014). Minimal processing of fresh fruit, vegetables, and juices. In: Sun, D. W., (ed.), *Emerging Technologies for Food Processing* (2nd Edn., pp. 583–597). Elsevier: Kidlington, Oxford, UK.
121. Cantwell, M., & Suslow, T., (1999). In: *Fresh-Cut Fruits and Vegetables: Aspects of Physiology, Preparation and Handling That Affect Quality* (pp. 1–22).
122. Dhall, R. K., (2013). Advances in edible coatings for fresh fruits and vegetables: A review. *Crit. Rev. Food. Sci. Nutr.*, *53*(5), 435–450.
123. Cisneros-Zevallos, L., & Krochta, J. M., (2002). Internal modified atmospheres of coated fresh fruits and vegetables: Understanding relative humidity effects. *J. Food Sci.*, *67*(6), 1990–1995.
124. Olivas, G. I. I., & Barbosa-Cánovas, G., (2009). Edible films and coatings for fruits and vegetables. In: Huber, K. C., & Embuscado, M. E., (eds.), *Edible Films and Coatings for Food Applications* (pp. 211–244). Springer: New York, NY, USA.
125. McHugh, T. H., & Krochta, J. M., (1994). Sorbitol- vs glycerol-plasticized whey protein edible films: Integrated oxygen permeability and tensile property evaluation. *J. Agric. Food Chem.*, *42*(4), 841–845.

126. Mellinas, C., Valdés, A., Ramos, M., Burgos, N., Garrigós, M. D. C., & Jiménez, A., (2016). Active edible films: Current state and future trends. *J. Appl. Polym. Sci.*, *133*(2), n/a-n/a.
127. Montero-Calderon, M., Soliva-Fortuny, R., & Martín-Belloso, O., (2016). Edible packaging for fruits and vegetables. In: Ribeiro, C. M. Â. P., Correia, P. R. N., Da Silva, R. Ó. L., Couto, T. J. A., & Augusto, V. A., (eds.), *Edible Food Packaging: Materials and Processing Technologies* (Vol. 36, pp. 353–382). CRC Press: Boca Raton, FL, USA.
128. Morillon, V., Debeaufort, F., Blond, G., Capelle, M., & Voilley, A., (2002). Factors affecting the moisture permeability of lipid-based edible films: A review. *Crit. Rev. Food Sci. Nutr.*, *42*(1), 67–89.
129. Vargas, M., Pastor, C., Chiralt, A., McClements, D. J., & González-Martínez, C., (2008). Recent advances in edible coatings for fresh and minimally processed fruits. *Crit. Rev. Food Sci. Nutr.*, *48*(6), 496–511.
130. Campos, C. A., Gerschenson, L. N., & Flores, S. K., (2011). Development of an edible films and coatings with antimicrobial activity. *Food Bioproc. Tech.*, *4*(6), 849–875.
131. Ribeiro, C., Vicente, A. A., Teixeira, J. A., & Miranda, C., (2007). Optimization of an edible coating composition to retard strawberry fruit senescence. *Postharvest Biol. Technol.*, *44*(1), 63–70.
132. García, M. A., Martino, M. N., & Zaritzky, N. E., (1998). Plasticized starch-based coatings to improve strawberry (*Fragaria × Ananassa*) quality and stability. *J. Agric. Food Chem.*, *46*(9), 3758–3767.
133. Kittur, F., Saroja, N., & Habibunnisa, T. R., (2001). Polysaccharide-based composite coating formulations for shelf life extension of fresh banana and mango. *Eur. Food Res. Technol.*, *213*(4), 306–311.
134. Brasil, I. M., Gomes, C., Puerta-Gomez, A., Castell-Perez, M. E., & Moreira, R. G., (2012). Polysaccharide-based multilayered antimicrobial edible coating enhances quality of fresh-cut papaya. *LWT-Food Sci. Technol.*, *47*(1), 39–45.
135. Cisneros-Zevallos, L., & Krochta, J. M., (2003). Whey protein coatings for fresh fruits and relative humidity effects. *J. Food Sci.*, *68*(1), 176–181.
136. Soazo, M., Pérez, L. M., Rubiolo, A. C., & Verdini, R. A., (2015). Prefreezing application of whey protein-based edible coating to maintain quality attributes of strawberries. *Int. J. Food Sci. Technol.*, *50*(3), 605–611.
137. Vargas, M., Albors, A., Chiralt, A., & González-Martínez, C., (2006). Quality of cold-stored strawberries as affected by chitosan–oleic acid edible coatings. *Postharvest Biol. Technol.*, *41*(2), 164–171.
138. Abugoch, L., Tapia, C., Plasencia, D., Pastor, A., Castro-Mandujano, O., López, L., et al., (2016). Shelf life of fresh blueberries coated with quinoa protein/chitosan/sunflower oil edible film. *J. Sci. Food Agric.*, *96*(2), 619–626.
139. Artés, F., Gómez, P. A., & Artés-Hernández, F., (2007). Physical, physiological and microbial deterioration of minimally fresh processed fruits and vegetables. *Food Sci. Technol. Int.*, *13*(3), 177–188.
140. Lee, J. Y., Park, H. J., Lee, C. Y., & Choi, W. Y., (2003). Extending shelf life of minimally processed apples with edible coatings and antibrowning agents. *LWT-Food Sci. Technol.*, *36*(3), 323–329.

141. Pasha, I., Saeed, F., Sultan, M. T., Khan, M. R., & Rohi, M., (2014). Recent developments in minimal processing: A tool to retain nutritional quality of food. *Crit. Rev. Food Sci. Nutr.*, *54*(3), 340–351.
142. Chandra Mohan, C., Radha krishnan, K., Babuskin, S., Sudharsan, K., Aafrin, V., Lalitha priya, U., et al., (2017). Active compound diffusivity of particle size reduced *S. aromaticum* and *C. cassia* fused starch edible films and the shelf life of mutton (*Capra aegagrus hircus*) meat. *Meat Sci.*, *128*, 47–59.
143. Hall, D. J., (2011). Edible coatings from lipids, waxes, and resins. In: Baldwin, E. A., Hagenmaier, R., & Bai, J., (eds.), *Edible Coatings and Films to Improve Food Quality* (2nd edn., pp. 79–101). CRC Press: Boca Raton, FL, USA.
144. Valencia-Chamorro, S. A., Palou, L., Del Río, M. A., & Pérez-Gago, M. B., (2011). Antimicrobial edible films and coatings for fresh and minimally processed fruits and vegetables: A review. *Crit. Rev. Food. Sci. Nutr.*, *51*(9), 872–900.
145. Das, D. K., Dutta, H., & Mahanta, C. L., (2013). Development of a rice starch-based coating with antioxidant and microbe-barrier properties and study of its effect on tomatoes stored at room temperature. *LWT-Food Sci. Technol.*, *50*(1), 272–278.
146. Guerreiro, A. C., Gago, C. M. L., Faleiro, M. L., Miguel, M. G. C., & Antunes, M. D. C., (2015). The effect of alginate-based edible coatings enriched with essential oils constituents on Arbutus unedo L. fresh fruit storage. *Postharvest Biol. Technol.*, *100*, 226–233.
147. González-Estrada, R. R., Chalier, P., Ragazzo-Sánchez, J. A., Konuk, D., & Calderón-Santoyo, M., (2017). Antimicrobial soy protein based coatings: Application to persian lime (citrus Latifolia Tanaka) for protection and preservation. *Postharvest Biol. Technol.*, *132*, 138–144.
148. Lee, J. Y., Park, H. J., Lee, C. Y., & Choi, W. Y., (2003). Extending shelf life of minimally processed apples with edible coatings and antibrowning agents. *LWT – Food Science and Technology*, *36*(3), 323–329.
149. Brasil, I. M., Gomes, C., Puerta-Gomez, A., Castell-Perez, M. E., & Moreira, R. G., (2012). Polysaccharide-based multilayered antimicrobial edible coating enhances quality of fresh-cut papaya. *LWT – Food Science and Technology*, *47*(1), 39–45.
150. Rossi, M. G., Di Pierro, P., Mariniello, L., Esposito, M., Giosafatto, C. V. L., & Porta, R., (2017). Fresh-cut fruit and vegetable coatings by transglutaminase-crosslinked whey protein/pectin edible films. *LWT-Food Sci. Technol.*, *75*, 124–130.
151. Han, J. H., (2014). A review of food packaging technologies and innovations. In: Han, J. H., (ed.), *Innovations in Food Packaging*(2nd edn.,pp. 3–12). Academic Press: San Diego, CA, USA.
152. Cutter, C. N., (2006). Opportunities for bio-based packaging technologies to improve the quality and safety of fresh and further processed muscle foods. *Meat Sci.*, *74*(1), 131–142.
153. Benjakul, S., Nagarajan, M., & Prodpan, T., (2016). Films and coatings from collagen and gelatin. In: García, M. P. M., Gómez-Guillén, M. C., López-Caballero, M. E.,& Barbosa-Cánovas, G. V., (eds.), *Edible Films and Coatings Fundamentals and Applications* (pp. 103–124). CRC Press: Boca Raton, FL.
154. Antoniewski, M. N., & Barringer, S. A., (2010). Meat shelf life and extension using collagen/gelatin coatings: A review. *Crit. Rev. Food Sci. Nutr.*, *50*(7), 644–653.

155. Ou, C. Y., Tsay, S. F., Lai, C. H., & Weng, Y. M., (2002). Using gelatin-based antimicrobial edible coating to prolong shelf life of tilapia fillets. *J. Food Qual., 25*(3), 213–222.
156. Gómez-Estaca, J., Bravo, L., Gómez-Guillén, M. C., Alemán, A., & Montero, P., (2009). Antioxidant properties of tuna-skin and bovine-hide gelatin films induced by the addition of oregano and rosemary extracts. *Food Chem., 112*(1), 18–25.
157. Ahmed, I., Lin, H., Zou, L., Brody, A. L., Li, Z., Qazi, I. M., et al., (2017). A comprehensive review on the application of active packaging technologies to muscle foods. *Food Control, 82*, 163–178.
158. Antoniewski, M. N., Barringer, S. A., Knipe, C. L., & Zerby, H. N., (2007). Effect of a gelatin coating on the shelf life of fresh meat. *J. Food Sci., 72*(6), 382–387.
159. Joukar, F., Hosseini, S. M. H., Moosavi-Nasab, M., Mesbahi, G. R., & Behzadnia, A., (2017). Effect of Farsi gum-based antimicrobial adhesive coatings on the refrigeration shelf life of rainbow trout fillets. *LWT-Food Sci. Technol., 80*, 1–9.
160. Ojagh, S. M., Núñez-Flores, R., López-Caballero, M. E., Montero, M. P., & Gómez-Guillén, M. C., (2011). Lessening of high-pressure-induced changes in Atlantic salmon muscle by the combined use of a fish gelatin–lignin film. *Food Chem., 125*(2), 595–606.
161. Volpe, M. G., Siano, F., Paolucci, M., Sacco, A., Sorrentino, A., Malinconico, M., & Varricchio, E., (2015). Active edible coating effectiveness in shelf life enhancement of trout (*Oncorhynchusmykiss*) fillets. *LWT-Food Sci. Technol., 60*(1), 615–622.
162. Martin, X. K. A., Hauzoukim, K. N., Balange, A. K., Chouksey, M. K., & Gudipati, V., (2017). Functionality of chitosan in batter formulations for coating of fish sticks: Effect on physicochemical quality. *Carbohydr. Polym., 169*, 433–440.
163. Gómez-Guillén, M. C., Pérez-Mateos, M., Gómez-Estaca, J., López-Caballero, E., Giménez, B., & Montero, P., (2009). Fish gelatin: A renewable material for developing active biodegradable films. *Trends Food Sci. Technol., 20*(1), 3–16.
164. Biji, K. B., Ravishankar, C. N., Mohan, C. O., & Srinivasa, G. T. K., (2015). Smart packaging systems for food applications: A review. *J. Food Sci. Technol., 52*(10), 6125–6135.
165. Gómez-Estaca, J., Montero, P., Giménez, B., & Gómez-Guillén, M. C., (2007). Effect of functional edible films and high pressure processing on microbial and oxidative spoilage in cold-smoked sardine (*Sardina pilchardus*). *Food Chem., 105*(2), 511–520.
166. Matiacevich, S., Acevedo, N., & López, D., (2015). Characterization of an edible active coating based on alginate–thyme oil–propionic acid for the preservation of fresh chicken breast fillets. *J. Food Process. Preserv., 39*(6), 2792–2801.
167. Ravishankar, S., Zhu, L., Olsen, C. W., McHugh, T. H., & Friedman, M., (2009). Edible apple film wraps containing plant antimicrobials inactivate foodborne pathogens on meat and poultry products. *J. Food Sci., 74*(8), 440–445.
168. Feng, X., Ng, V. K., Mikš-Krajnik, M., & Yang, H., (2017). Effects of fish gelatin and tea polyphenol coating on the spoilage and degradation of myofibril in fish fillet during cold storage. *Food Bioproc. Tech., 10*(1), 89–102.
169. Ozogul, Y., Yuvka, İ., Ucar, Y., Durmus, M., Kösker, A. R., Öz, M., et al., (2017). Evaluation of effects of nanoemulsion based on herb essential oils (rosemary, laurel, thyme and sage) on sensory, chemical and microbiological quality of rainbow

trout (*Oncorhynchus mykiss*) fillets during ice storage. *LWT-Food Sci. Technol.*, 75, 677–684.
170. Alsaggaf, M. S., Moussa, S. H., & Tayel, A. A., (2017). Application of fungal chitosan incorporated with pomegranate peel extract as edible coating for microbiological, chemical and sensorial quality enhancement of Nile tilapia fillets. *Int. J. Biol. Macromolec.*, 99, 499–505.
171. Zhong, Y., Cavender, G., & Zhao, Y., (2014). Investigation of different coating application methods on the performance of an edible coatings on Mozzarella cheese. *LWT-Food Sci. Technol.*, 56(1), 1–8.
172. Ustunol, Z., (2009). Edible films and coatings for meat and poultry. In: Huber, K. C.,& Embuscado, M. E., (eds.), *Edible Films and Coatings for Food Applications*(pp. 245–268). Springer: New York, NY.
173. De Castro, R. J. S., Domingues, M. A. F., Ohara, A., Okuro, P. K., Dos Santos, J. G., Brexó, R. P., et al., (2017). Whey protein as a key component in food systems: Physicochemical properties, production technologies and applications. *Food Structure.* 14, 17–29.
174. Andrade, R. D., Skurtys, O., & Osorio, F. A., (2012). Atomizing spray systems for application of an edible coatings. *Compr. Rev. Food Sci. Food Saf.*, 11(3), 323–337.
175. Gomez, R., Carmona, M. A., & Fernández-Salguero, J., (1991). Estudio de los alimentos de humedad intermedia españoles. I. Actividad de agua y pH. *II Jornadas Científicas Sobre "Alimentación Española,"* 123–130.
176. Janes, M. E., & Dai, Y., (2012). Edible films for meat, poultry and seafood. In: Kerry, J. P., (ed.), *Advances in Meat, Poultry and Seafood Packaging* (pp. 504–521). Woodhead Publishing: Cambridge, UK.
177. Kapetanakou, A. E., Karyotis, D., & Skandamis, P. N., (2016). Control of *Listeria monocytogenes* by applying ethanol-based antimicrobial edible films on ham slices and microwave-reheated frankfurters. *Food Microbiol.*, 54, 80–90.
178. Gómez-Guillén, M. C., Ihl, M., Bifani, V., Silva, A., & Montero, P., (2007). Edible films made from tuna-fish gelatin with antioxidant extracts of two different murta ecotypes leaves (*Ugni molinae* Turcz). *Food Hydrocoll.*, 21(7), 1133–1143.
179. Ollé Resa, C. P., Gerschenson, L. N., & Jagus, R. J., (2016). Starch edible film supporting natamycin and nisin for improving microbiological stability of refrigerated Argentinian Port Salut cheese. *Food Control*, 59, 737–742.
180. Fajardo, P., Martins, J. T., Fuciños, C., Pastrana, L., Teixeira, J. A., & Vicente, A. A., (2010). Evaluation of a chitosan-based edible film as the carriers of natamycin to improve the storability of Saloio cheese. *J. Food Eng., 101*(4), 349–356.
181. Pinto, A. M. B., Santos, T. M., Caceres, C. A., Lima, J. R., Ito, E. N., & Azeredo, H. M. C., (2015). Starch-cashew tree gum nanocomposite films and their application for coating cashew nuts. *LWT-Food Sci. Technol.*, 62(1), 549–554.
182. Atarés, L., Pérez-Masiá, R., & Chiralt, A., (2011). The role of some antioxidants in the HPMC film properties and lipid protection in coated toasted almonds. *J. Food Eng., 104*(4), 649–656.
183. Soukoulis, C., Yonekura, L., Gan, H. H., Behboudi-Jobbehdar, S., Parmenter, C., & Fisk, I., (2014). Probiotic edible films as a new strategy for developing functional bakery products: The case of pan bread. *Food Hydrocoll.*, 39, 231–242.

184. Liang, J., Yan, H., Zhang, J., Dai, W., Gao, X., Zhou, Y., Wan, X., & Puligundla, P., (2017). Preparation and characterization of antioxidant edible chitosan films incorporated with epigallocatechin gallate nanocapsules. *Carbohydr. Polym.*, *171*, 300–306.
185. Sabaghi, M., Maghsoudlou, Y., Khomeiri, M., & Ziaiifar, A. M., (2015). Active edible coating from chitosan incorporating green tea extract as an antioxidant and antifungal on fresh walnut kernel. *Postharvest Biol. Technol.*, *110*, 224–228.
186. Lee, H., Kim, J. E., & Min, S. C., (2015). Quantitative risk assessments of the effect of an edible defatted soybean meal-based antimicrobial film on the survival of *Salmonella* on ham. *J. Food Eng.*, *158*, 30–38.
187. Lekjing, S., (2016). A chitosan-based coating with or without clove oil extends the shelf life of cooked pork sausages in refrigerated storage. *Meat Sci.*, *111*, 192–197.
188. Ferreira, S. L. E., Naponucena, L. D. O. M., Da Silva Santos, V., Silva, R. P. D., De Souza, C. O., Evelyn, G. L. S. I., et al., (2016). Development and application of an edible film of active potato starch to extend mini panettone shelf life. *LWT-Food Sci. Technol.*, *73*, 311–319.
189. Catarino, M. D., Alves-Silva, J. M., Fernandes, R. P., Gonçalves, M. J., Salgueiro, L. R., Henriques, M. F., et al., (2017). Development and performance of whey protein active coatings with *Origanum virens* essential oils in the quality and shelf life improvement of processed meat products. *Food Control*, *80*, 273–280.
190. Upadhyay, A., Upadhyaya, I., Karumathil, D. P., Yin, H. B., Nair, M. S., Bhattaram, V., et al., (2015). Control of Listeria monocytogenes on skinless frankfurters by coating with phytochemicals. *LWT-Food Sci. Technol.*, *63*(1), 37–42.
191. Zhao, Y., Abbar, S., Phillips, T. W., Williams, J. B., Smith, B. S., & Schilling, M. W., (2016). Developing food-grade coatings for dry-cured hams to protect against ham mite infestation. *Meat Sci.*, *113*, 73–79.
192. Jridi, M., Abdelhedi, O., Souissi, N., Kammoun, M., Nasri, M., & Ayadi, M. A., (2015). Improvement of the physicochemical, textural and sensory properties of meat sausage by edible cuttlefish gelatin addition. *Food Biosci.*, *12*, 67–72.
193. Debeaufort, F., & Voilley, A., (2009). Lipid-based edible films and coatings. In: Huber, K. C., & Embuscado, M. E., (eds.), *Edible Films and Coatings for Food Applications* (pp. 135–168). Springer: New York, NY, USA.
194. Dangaran, K., Tomasula, P. M., & Qi, P., (2009). Structure and function of protein-based edible films and coatings. In: Huber, K. C., & Embuscado, M. E., (eds.), *Edible Films and Coatings for Food Applications* (pp. 25–56). Springer: New York, NY.

CHAPTER 2

ENCAPSULATION OF PREBIOTICS AND PROBIOTICS: A PROMISING ALTERNATIVE IN MODERN FOOD TECHNOLOGY

DIANA B. MUÑIZ-MÁRQUEZ,[1] JORGE E. WONG-PAZ,[1]
SILVIA M. GONZÁLEZ,[2] PEDRO AGUILAR-ZÁRATE,[1]
and ORLANDO DE LA ROSA[3]

[1] *Engineering Department, Technological Institute of Ciudad Valles, National Technological of México, Ciudad Valles, 79010, San Luis Potosí, Mexico, Tel. +52 (481) 38 1 20 44, E-mail: diana.marquez@tecvalles.mx*

[2] *Department of Chemical Engineering and Biochemistry, Technological Institute of Durango, National Technological of México, Durango, 34080, Durango, Mexico*

[3] *Food Research Department, School of Chemistry, Autonomous University of Coahuila, Saltillo, 25280, Coahuila, Mexico*

ABSTRACT

In recent years, the food industries have opted for the application of new technologies in the development of functional foods, such as encapsulation process, with the purpose of protect the bioactive compounds called prebiotics and probiotic during passage through the gastrointestinal zone and thereby ensure their total integrity and the viability during food product formulation and their storage. Encapsulation is a promising method because it allows the stabilization of prebiotics and probiotics by resistance to stomach acid pH and therefore offering health beneficial effects inside the digestive system. In addition, in this technique, it is possible the incorporation of materials from natural sources such as potato and maize starch, pectin and others natural products that act as encapsulating agents showing an excellent biocompatibility, biodegradability, nontoxicity, cost-effectiveness and therapeutic properties.

In this context, the aim of this chapter discusses the importance of the encapsulation process of prebiotic compounds and probiotics microorganism for the development of functional foods. As well as, bioengineering aspects chemical and physical characteristics of encapsulation of functional ingredients and some industrial applications.

2.1 INTRODUCTION

The food industry has opted for the application and use of the new technologies capable of protecting the functional properties of food products before and after consumption with the purpose of assuring the health of the consumer. In this sense, the encapsulation process is a novel alternative to preventing the deterioration of bioactive ingredients and therefore, this method is an effective approach for the incorporation of bioactive ingredients in functional foods and beverages [1]. The process of encapsulation of prebiotics or probiotics is through of one or two polymers such as alginate, chitosan, carrageenan, pectin and other polysaccharides. Encapsulation systems have the advantages of resisting the stomach acidity and fermentation by intestinal microbiota thus offering benefits to consumers of encapsulated products [2]. Many investigations have focused on these issues for example, in a study, modified maize starches were used for the encapsulation of probiotic microorganism (*L. plantarum* 299v) [3]. More recently, Damodharan et al. [4] studied the effect of microcapsules of alginate to fenugreek gum in the viability of probiotics microorganism such as *Pediococcus pentosaceus, Lactobacillus plantarum, Lactobacillus fermentum* and *Lactobacillus helveticus*. They observed that the combination of alginate-fenugreek gum locust bean gum matrix was the best for the encapsulation of bacteria probiotic because tolerated digestive condition in comparison with the non-encapsulated bacteria. Iravani et al. [67] mentioned that the selection of the best method plays an important role in the cell viability of the probiotics, in addition, are necessary high concentrations of bacterial population to assure their therapeutic properties of the foods before and after consumption. On the other hand, a new trend in the functional foods is the formulation of products enriched with symbiotic especially in the formulas for infant [5].

Therefore, this chapter focuses on the encapsulation processes of pre- and probiotics. Presents some advances of these systems as well as future prospects, advantages, and limitations of these techniques in the food industry.

2.2 GENERALITIES OF PREBIOTICS AND PROBIOTICS

2.2.1 PREBIOTICS

Prebiotics are defined as non-digestible food ingredients that beneficially affect the host selectively by stimulating the growth and activity of a limited number of bacteria and improving host health [6]. Prebiotics must fulfill the following criteria:

a) No hydrolysis or absorption in the upper part of the digestive system (mouth, stomach, small intestine).
b) Selective substrate for one or more desired bacteria species in the colon and stimulation of that species regarding growth and activation.
c) Able to positively influence the numeric proportion of different bacteria species in the colon.

The definition of prebiotics was later updated in 2004 and confirmed in the 6th Meeting of the International Scientific Association of Probiotics and Prebiotics (ISAPP) in 2008 and is defined as follows: "selectively fermented ingredients that allow specific changes, both in the composition and/or activity in the gastrointestinal microbiota that confers benefits upon host well-being and health" [7, 8]. According to the definition, prebiotics should provide benefits not only in the colon but in the whole gastrointestinal tract.

Several oligosaccharides are therefore generally accepted as prebiotics like inulin, fructooligosaccharides (FOS), galactooligosaccharides (GOS) and lactulose [9], but recently others have been proposed as prebiotics such as resistant starches, pectin, arabinoxylan, whole grains and non-carbohydrate compounds such as polyphenols, according to the authors wider definition of prebiotic "a prebiotic is a non-digestible compound that, through its metabolization by microorganisms in the gut, modulates composition and/or activity of the gut microbiota, thus conferring a beneficial physiological effect on the host" [10, 11].

2.2.2 PROBIOTICS

Probiotics are defined as live microorganisms which, when administered in adequate amounts, have a beneficial effect on the health of the host [12]. Not every fermented food containing live organisms is considered a probiotic, because to be considered in this classification the amount of these microorganisms should be specified and its effects have to be specifically studied [9]. To achieve beneficial effects in the host, probiotics do not need to colonize the target organ, but it should be reached by a sufficient number of live microorganisms so as to modify the microbiota and metabolism [9].

Thus, most probiotic strains must show good viability after passing through the upper gastrointestinal tract, and it could depend on many factors: Firstly, intrinsic probiotic factors and host-dependent factors such as, stomach acidity, time of acid exposure, the concentration and duration of exposure to bile salts, and others [13].

To be considered a probiotic, there should be enough evidence of conducted studies in humans showing the specific health benefits of specific strains. An example of some probiotics used in clinical trials are: *Saccharomyces boulardii*, *Lactobacillus rhamnosus* GG, *Bifidobacterium bifidum*, *Lactobacillus plantarum*, *Lactobacillus sporogens*, *Enterococcus* SF68, *Bifidobacterium lactis* BB12 (L), *Lactobacillus reuteri*, *Lactobacillus casei* (L), *Bifidobacterium longum* BB 536 (L), *Lactobacillus acidophilus* LA1, *Escherichia coli* Nissle 1917 (serotype 06:K5:H1) [9, 14].

2.3 BENEFITS OF PREBIOTICS AND PROBIOTICS

Great efforts in developing strategies in the daily diet for modulating the composition and activity of the microbiota, using prebiotics, probiotics, and a combination of both (symbiotic) has been explored [15–17, 66]. Given that prebiotics and probiotics act consequently and the main benefits will be listed generally for both.

2.3.1 ACTION MECHANISMS

According to the consensus of the International Scientific Association for Probiotics and Prebiotics [18] the mechanisms of action for probiotics can be classified in categories: colonization resistance, normalization of perturbed microbiota, production of short chain fatty acids and acidification of the medium, increased recovery of enterocytes, regulation of gastrointestinal transit, competitive exclusion of putrefactive or pathogenic bacteria. Mechanisms frequently present in probiotics: vitamin synthesis, bile salt metabolism, direct antagonism, enzymatic activity, reinforcement of intestinal barrier, neutralization of carcinogens. Mechanisms found only in specific strains: neurological effects, immunological effects, endocrinological effects and production of specific bioactive substances.

According to the mechanisms of action presented by prebiotics they can directly influence some important diseases common in the human population [19–21].

2.3.2 ROLE IN OBESITY

It is not clear that manipulating the bifidobacterial populations in the gut will have an impact on obesity, but there is promising evidence suggesting that prebiotics might have an impact on appetite, thus indirectly impacting upon weight gain.

Verhoef et al. [22] fed the prebiotic oligofructose (OF) to 28 healthy adults for 13 days and studied appetite profiles, energy intake and expression of the gut hormones PYY and GLP-1. They found that although oligofructose consumption did not suppress appetite, energy consumption was reduced by 11% on day 13 when consuming 16 g OF per day. Here, the expression of both gut hormones was increased. The authors suggest that this might be due to the production of elevated levels of SCFA by fermentation of OF [20].

A recent systematic review has been conducted on the effect of OF and inulin on appetite regulation, energy intake and weight loss in children and adults [23]. Studies amplifying this have been published by Cani et al. [24, 25] who showed that fructan prebiotics could influence satiety in humans. Similarly, Parnell and Reimer [26] demonstrated that the same type of prebiotics could influence hormonal regulation and therefore appetite in overweight humans. The conclusion is that OF and inulin may have a contribution to make to reducing energy intake and weight loss.

2.3.3 PREVENTION OF COLON CANCER

As said before prebiotics such as fructans allows quantitative and/or qualitative alterations of the beneficial microflora, which are able to inactivate mutagenic/carcinogenic/genotoxic compounds [27]. *Bifidobacterium* (BF) longum and lactulose increased the activity of colonic glutathione S-transferase (involved in the detoxification of toxic metabolites and carcinogens) [28].

Prebiotic fermentation leads to the production of SCFA such as butyrate, which has immunomodulation properties results include upregulation of the apoptosis. Butyrate is a preferred energy source for the mucosal cells which might inhibit neoplastic changes in cancer cells [29].

Another mechanism against colon cancer unchained by the proliferation of probiotic bacteria is the stimulation of the host's anti-tumor immunity [30–33, 68], by stabilizing epithelial tight junctions, inducing epithelial defensin production, inducing the anti-inflammatory and immunomodulatory capacity of T regulatory cells and of dendritic cells [34], stimulating B cells and natural killer cells, with role in controlling tumor promotion and progression [35].

2.3.4 GASTROINTESTINAL DISEASES

Travelers' Diarrhea is a common disease, presented most of the time due to bacterial infection, nevertheless is a health issue that can be prevented with the use of probiotics. A reported meta-analysis [36, 37] showed that some probiotics are effective in preventing travelers' diarrhea and estimated that up to 85% of these cases could be prevented with probiotics. The probiotic *Lactobacillus reuteri*, has proved to be effective inhibiting the colonization of the human gastric mucosa by *Helicobacter pylori* [38], this probiotic among others have shown to help to eradicate *H. pylori* during medical treatment [39].

A work with patients with active ulcerative colitis (UC) treated with 7.5 g/day or 15 g/day of inulin-type b-fructans for 9 weeks showed that the prebiotics shifted the fecal microbiota towards a FOS-induced co-metabolism between *Bifidobacterium* sp. and *Roseburia* sp., and showed significant increase in the level of total SCFA, butyrate and acetate production and significant reduction in the disease activity in the UC patients with a dose of 15 g/day [11, 69].

2.4 ENCAPSULATION PROCESSES

Prebiotics are important food components that confer health benefits when they are ingested in sufficient concentrations. However, has been a challenge to protect them during food processing. Encapsulation is defined as a method in which the bioactive ingredients are surrounded by a polymer coating to give capsules with many functional properties [40]. Is a method based in packing solid, which release their components upon applying certain conditions, for example, heating, diffusion and pressure [41]. In encapsulation techniques, particles with micrometric size are obtained (1–1000 μm). Therefore, these processes have been used by the formulation of cosmetics, pharmaceutical and food products.

Some studies have showed the efficacy of the prebiotics and probiotics encapsulation process, even higher when prebiotics and probiotics are encapsulated together. According to Peredo et al. [2], the effect of different prebiotics (potato starch, *Plantago psyllium* and inulin) co-encapsulated with alginate with the purpose of observing the viability of *Lactobacillus casei* Shirota and *Lactobacillus plantarum*. These researchers report that with the encapsulation process was possible the obtaining of spherical capsules with a great during their storage. In addition, capsules of *Plantago psyllium* were the best because offered a good protection in the intestinal zone.

Another example, Sathyabama et al. [42] evaluated the behavior of two probiotic microorganisms *Sthapylococcus succinus* and *Enterococcus fecium* which were co-encapsulated with natural prebiotic (sugar beet and chicory) and alginate. The results showed that improvement in viability of co-encapsulated microorganism in acid and bile conditions (pH 2–3). Zuidam and Shimoni [41] mentioned that according with the International Standards (International Dairy Federation) the probiotic ingredients contain a minimum of 10^7 viable probiotic microorganism per gram of product or 10^9 bacteria per serving size when sold equivalent to 10^{6-8}

cells/g feces. But this is sometimes not possible because probiotic bacteria death occurs when the functional foods are stored at high temperatures. Therefore, has been necessary to find technological and economic strategies for the protection of probiotics during the time of processing until their consumption. Encapsulation methods include *spray drying, extrusion, spray chilling, cocrystallization, coacervation* and *freeze drying*. Spray drying is the most commonly used technique of encapsulation and preservation for foods. Onwulata [43] mentioned that spray drying is a method inexpensive and simple. This technique involves the formation of a matrix layer with the encapsulating material and forcing materials into the matrix by a spinning atomizer resulting in the formation of spheres. Spray drying is one of the procedures most used in the food industry because this allows to obtain high-quality products with little water activity which facilitates their transport and storage. This method has the advantage of obtaining particles spherical and small with homogeneous distribution [40].

Encapsulation process have some advantages on the bioactive compounds such as to provide a protective barrier between them and the destructive factors prevalent in the surrounding environment such as heat, oxygen, and moisture [44]. Two main consideration in the encapsulation process might be take into account: (i) keeping them alive until they reach the target site, and (ii) effective release of the entrapped bioactive compound. In this sense, the material and the size of the capsule are an important factor for the stability and efficacy of the encapsulated compound [45], if this consideration is not taken into account, some limitations of the encapsulation process could appear. A high capsule could result in a limit releasing of the bioactive compounds, in contrast, a small capsule could be not enough to protect the bioactive compounds during the whole process of food preparation and digestion. Therefore, always is necessary to study a specific material to encapsulate with a specific bioactive compound in a specific process.

Mentioning a specific case, the objective of the probiotic encapsulation is to protect the microorganism whilst passing upper gastrointestinal tract until the bowel of human, but limiting the use of the prebiotics and probiotics directly in the food. Because also, probiotics could be useful to control the pathogens in the food in a non-encapsulated form, but the viability of them could be at risky because of non-encapsulation to protect of the gastrointestinal conditions would be present. In any case, an alternative could be to use encapsulated probiotics to ensure their viability in the human bowel and, at the same time to use non-encapsulated probiotics to control the pathogenic microorganisms directly in the food.

2.5 MATERIALS FOR THE ENCAPSULATION

Selection of appropriate coating carriers and the encapsulation method influence type of capsule formed and therefore in the rate of retention and biofunctionality of the compounds. Is necessary to know the physicochemical properties of food grade materials employed in the encapsulation considering the interactions of molecules such as lipids, carbohydrates, proteins and other mixtures [41]. In addition, when microcapsules with probiotics are produced, it is of special interest to consider the factors that may affect them, probiotics are sensitive to pH changes, digestive enzymes, mechanical stress and transport among others, so that the materials used as encapsulating agents are an important element. Then, the material used in the encapsulation might have several advantages such as non-toxic, easy to handle, bioavailable, biocompatible, low cost and easy bead formation by ionotropic gelation, increasing probiotic survival, stability at low pH, swell in weakly basic solutions, resistance to low and high temperatures and osmotic pressure, and their incorporation without affecting food sensory quality [4]. Table 2.1 shows some of the most used material for encapsulation.

TABLE 2.1 Natural Sources Commonly Used as Encapsulates in the Food Industry

Polysaccharides	Proteins	Lipids and waxes
Starches	Sodium caseinate	Vegetable and hydrogenated fats
Maltodextrins	Proteins from whey, wheat and soy	Palm stearin
Syrups and Gums	Gelatins	Carnauba and Bees wax
Pectin		Polyethylene glycol
Alginates		Vegetable oils
Carrageenan		Others
Chitosan		
Celluloses		

From: Vieira da Silva et al., [46].

2.5.1 POLYSACCHARIDES

Polysaccharides are used as food supplement due to that are one of the most versatile biopolymers produced by nature. Their versatility is according to hydrophobic and hydrophilic motifs with variable extension in the same polymer that provides matrices sensitive to environmental conditions [47].

2.5.2 PROTEINS

Proteins from animal sources (casein, eggs, gelatin and whey) are widely employed for the nanoparticle formation. The selection of a suitable protein for the application depends on certain characteristics such as permeability, size, charge and degradability which impact to some properties of the bioactive components to be encapsulated [47].

2.5.3 LIPIDS

The lipidic moieties have a hydrophilic head and hydrophobic acyl chain in addition have important physicochemical properties useful for the combination of various functional groups of a many varieties of molecules. Purification process in lipids is a fundamental step for the determining the physicochemical characteristics of the formulations [47].

2.6. ENCAPSULATED PREBIOTICS AND PROBIOTICS AS FUNCTIONAL FOODS IN THE FOOD INDUSTRY

As consumers have been conscious that their health and well-being depend largely on the type of food consumed in their diet, the demand in functional food has had an explosive increase. Upon this demand, the food industry faces the challenge of adapting and designing appropriate processing systems to preserve and improve functional, physicochemical and sensorial properties. Innovations that food industry needs to address include other fields as packaging and commercialization systems [48].

A resource to transform traditional food into functional food is its supplementation with bioactive compounds as probiotics, prebiotics, vitamins and fatty acids amongst other. To not affect the functionality and effectiveness of these components, it is necessary to guarantee its stability, viability and release. The application of diverse encapsulation techniques has contributed to achieve this goal [46]. A great number of food have been improved with probiotics and launched to the market as fermented and not fermented food, ice cream, juices fruit, butter peanut and cereal products; besides, the most used probiotics belong to the genera *Lactobacillus* and *Bifidobacterium* [49].

Probiotic bacteria are widely used within the food industry, and there is a great interest on the microencapsulation technique to add bacteria in food matrices; this technique has demonstrated to be effective to protect bacteria viability. Moumita et al. [50] investigated to evaluate the encapsulation efficiency on probiotic survival in a dry food matrix during the storage, the encapsulating material was a prebiotic extract from *Pleorotus ostreatus*. Bacteria from *Lactobacillus* genera were used, and they conclude that probiotic survival was favorable. Also, they report an additional protection against stress in the gastrointestinal tract.

In many beverages microcapsules are used for having as a purpose to get included in functional beverages, and attractive materials for their low cost are the corn syrup solids, nevertheless, they are not convenient for diabetics, and prebiotics as chicory inulin begin to emerge as another wall material option [51]. Table 2.2 shows some of the applications that have been carried out using prebiotics as wall material.

2.6.1 FOOD ADDED WITH ENCAPSULATED PROBIOTICS IN THE MARKET

There exists in the market a large variety of food in which microencapsulation or immobilization have been supplemented with probiotics (Table 2.3).

Besides the examples listed above, every day the diversity of non-dairy food enriched with probiotics continues to increase, such as dried fruits, carrot juice, mixed cereal beverage, vegetable-based drinks and fruit bars amongst other [59].

TABLE 2.2 Prebiotics as Encapsulating Agents for Probiotics

Encapsulating agents	Probiotic	Food	Application	Author
Inulin, GOS, alginate, chitosan	Lactobacillus acidophilus Lactobacillus casei	Yogurt Orange juice	Enhance probiotic viability.	Krasaekoopt & Watcharapoka, [52]
Inulin, glyceryl, dipalmitostearate	Bifidobacterium longum	Additive	Improve the protection of probiotics.	Amakiri et al., [53]
Eleutherine americana extract, FOS, alginate	Bifidobacterium longum	Fresh milk tofu Pineapple juice	Improvement of probiotic viability in gastrointestinal conditions and storage. Obtain sensory scores.	Phoem et al., [54]
Native inulin, cross-linked inulin, acetylated inulin, sodium caseinate		Alpha tocopherol	Alpha tocopherol protection.	García et al., [55]
Native inulin, short chain inulin, long chain inulin, alginate, chitosan	Lactobacillus casei	Additive	Improve the survival and stability of probiotic.	Darjani et al., [56]
Inulin, reconstituted skim milk	Bifidobacterium BB-12	Frozen yogurt	Viability of the probiotic. Chemical and rheological properties.	Pinto et al., [70]
Hi-Maize starch, alginate	Lactobacillus acidophilus Lactobacillus casei Bifidobacterium infantis	Yoghurt	Enhance the survival of the probiotic bacteria.	Sultana et al., [57]
Oligosaccharides extracted from sugar beet, chicory and oats, alginate	Staphyloccocus succinus Enterococcus fecium	Additive	Efficiency in improving the viability.	Sathyabama et al., [42]

Encapsulation of Prebiotics and Probiotics

TABLE 2.3 Example of Encapsulated Probiotic Used in Functional Foods

Immobilization/ encapsulation material	Probiotic bacteria	Functional food product
Alginate	*Bifidobacterium bifidum* or *B. infantis*	Mayonnaise
	Lactobacillus reuteri	Sausages
	Lactobacillus acidophilus BCRC 10695	Tomato juice
	Lactobacillus casei NCDC 298	Synbiotic milk chocolate
Apple derivates	*Lactobacillus casei* ATCC 393	Milk products
Pear derivates	*Lactobacillus casei* ATCC 393	Milk products
Chitosan and alginate	*Lactobacillus casei* 01 or *L. acidophilus* 547	Milk products
Whey protein	*Lactobacillus trhamnosus* R011	Frozen cranberry

Source: Mitropoulou et al., [58].

2.7 FUTURE PERSPECTIVES

The use of the encapsulation is not recent and it has been evolving as its application has spread to different areas, being one of the most important in the food industry. The potential in the food area has not been exploited in its totality, because new applications emerge day by day. Today, the need to search alternative materials that replace the traditional with sustainable materials from renewable resources has become pressing due to stricter international regulations [60]. These alternative materials might improve the characteristic of the traditional, such as resistance to strong pH changes, preservation of prebiotic and probiotic bioactivity, allowing the easy, controlled and selective releasing of them.

Some attempts are being performed, an example of these alternative materials are the cyclodextrins which are generally regarded as safe (GRAS), chemically and physically stable, with a hydrophobic interior and a hydrophilic exterior and they are not absorbed in the upper gastrointestinal tract, but are completely metabolized by the colon microflora [61].

The encapsulation of prebiotics and probiotics using a polymeric mixture as encapsulating continues to be explored. Co-encapsulation of lactic acid bacteria and prebiotic with alginate-fenugreek gum-locust bean gum matrix was recently studied by Damodharan et al. [4]. These authors reported this matrix showed a good encapsulation efficiency and dissolvability in colonic fluid compared with the non-encapsulated system.

Nanoscale materials are taking advantage area in encapsulation manufacturing because the materials in this scale are shown a lot of benefits mainly provided by a greater surface area and increased solubility, enhancing the characteristics above mentioned. Nanomaterials based in proteins are particularly interesting because they are relatively easy to prepare and their size distribution can be easily monitored [46].

Another novelty technique in the field of encapsulation of prebiotic and probiotics is the electrospinning. This technique uses that uses electrostatic forces to create polymer fibers. The absence of high temperatures during this process makes it very useful for encapsulating of thermolabile compounds and organisms [62]. The electrospinning importance is due to this methodology produce matrices in the micro and nano-range which are potentials as systems for sustained and controlled release [63]. Through this technique has already been demonstrated that using proteins and carbohydrates as the encapsulating materials the viability and stability of the probiotic bacteria and bacteriocins during food processing and storage are maintained [63, 65].

Finally, the use in foods of both non-encapsulated and encapsulated prebiotics and probiotics to show the synergistic effect require further studies to show the beneficial impact of these systems directly in the foods and consumers. The appropriate encapsulation process should be chosen based on the component to be encapsulated (prebiotic and probiotic, or both), the matrix (encapsulation material) and where it will be applied [64]. The development of encapsulated prebiotics and probiotics is an interesting topic in the food industry with a high expansion potential and brief it will be conventional part of the food industry process.

KEYWORDS

- encapsulation
- functional foods
- prebiotics
- probiotics

REFERENCES

1. Donsì, F., Sessa, M., Mediouni, H., Mgaidi, A., & Ferrari, G., (2011). Encapsulation of bioactive compounds in nanoemulsion-based delivery systems. *Procedia Food Science, 1*, 1666–1671.
2. Peredo, A. G., Beristain, C. I., Pascual, L. A., Azuara, E., & Jimenez, M., (2016). The effect of prebiotics on the viability of encapsulated probiotic bacteria. *LWT – Food Science and Technology, 73*, 191–196.
3. Li, H.: Thuy Ho, V. T., Turner, M. S., & Dhital, S., (2016). Encapsulation of Lactobacillus plantarum in porous maize starch. *LWT-Food Science and Technology, 74*, 542–549.
4. Damodharan, K., Arunachalam Palaniyandi, S., Hwan Yang, S., & Won Suh, J. (2017). Co-encapsulation of lactic acid bacteria and prebiotic with alginate-fenugreek gum-locust bean gum matrix: Viability of encapsulated bacteria under simulated gastrointestinal condition and during storage time. *Biotechnology and Bioprocess Engineering, 22*, 265–271.
5. Mugambi, M. N., Young, T., & Blaauw, R., (2014). Application of evidence on probiotics, prebiotics and synbiotics by food industry: A descriptive study. *BMC Research Notes, 7*, 754.
6. Gibson, G. R., & Roberfroid, M. B., (1995). Dietary modulation of the human colonic microbiota: Introducing the concept of prebiotics. *The Journal of Nutrition, 125*, 1401–1412.
7. Gibson, G., Probert, H., Loo, J., Rastall, R., & Roberfroid, M., (2004). Dietary modulation of the human colonic microbiota: Updating the concept of prebiotics. *Nutrition Research Reviews, 17*, 259–275.
8. Gibson, G. R., Scott, K. P., Rastall, R. A., Touhy, K. M., Hotchkiss, A., Dubert-Ferrandon, A., et al., (2010). Dietary prebiotics: Current status and new definition. *Journal of Food Science and Technology Bulletin: Functional Foods, 7*, 1–19.
9. Olveira, G., & González-Molero, I., (2016). Actualización de probióticos, prebióticos y simbióticos en nutrición clínica. *Endocrinología y Nutrición., 63*, 482–494.
10. Bindels, L. B., Delzenne, N. M., Cani, P. D., & Walter, J., (2015). Towards a more comprehensive concept for prebiotics. *Nature Reviews Gastroenterology and Hepatology., 12*, 303–310.
11. Valcheva, R., & Dieleman, L. A., (2016). Prebiotics: Definition and protective mechanisms. Best Practice and Research: *Clinical Gastroenterology, 30*, 27–37.
12. Ghouri, Y. A., Richards, D. M., Rahimi, E. F., Krill, J. T., Jelinek, K. A., DuPont, A. W., (2014). Systematic review of randomized controlled trials of probiotics, prebiotics, and synbiotics in inflammatory bowel disease. *Clinical and Experimental Gastroenterology, 7*, 473–487.
13. Bezkorovainy, A., (2001). Probiotics: Determinants of survival and growth in the gut. *The American Journal of Clinical Nutrition, 73*, 399–405.
14. Organización Mundial de la Salud, (2011). Guía práctica de la organización mundial de gastroenterología: Probióticos y prebióticos. Guías Mundiales de la WGO Probióticos y prebióticos, p. 29.

15. Howlett, J., (2008). Functional foods-from science to health and claims. *ILSI Europe-Concise Monograph Series*, Brussels, Belgium. ISBN9789078637110, Ref. 18, pp. 36.
16. Szajewska, H., (2010). Probiotics and prebiotics in preterm infants: Where are we? Where are we going?. *Early Human Development, 86*, 81–86.
17. Dominguez, A., Rodrigues, L., Lima, N., & Teixeira, J., (2013). An overview of the recent developments on fructooligosaccharide production and applications. *Food Bioprocess and Technology, 7*, 324–337.
18. Hill, C., Guarner, F., Reid, G., Gibson, G. R., Merenstein, D. J., Pot, B., Morelli, L., et al., (2014). Expert consensus document: The international scientific association for probiotics and prebiotics consensus statement on the scope and appropriate use of the term probiotic. *Nature Reviews Gastroenterology and Hepatology, 11*, 506–514.
19. Scheid, M. M. A., Moreno, Y. M. F., Maróstica Junior, M. R., & Pastore, G. M., (2013). Effect of prebiotics on the health of the elderly. *Food Research International, 53*, 26–432.
20. Rastall, R. A., & Gibson, G. R., (2015). Recent developments in prebiotics to selectively impact beneficial microbes and promote intestinal health. *Current Opinion in Biotechnology, 32C*, 42–46.
21. Wu, R. Y., Määttänen, P., Napper, S., Scruten, E., Li, B., Koike, Y., et al., (2017). Nondigestible oligosaccharides directly regulate host kinome to modulate host inflammatory responses without alterations in the gut microbiota. *Microbiome., 5*, 135, 1–15.
22. Verhoef, S., Meyer, D., & Westerterp, K., (2011). Effects of oligofructose on appetite profile, glucagon-like peptide 1 and peptide YY3–36 concentrations and energy intake. *British Journal of Nutrition, 106*, 1757–1762.
23. Liber, A., & Szajewska, H., (2013). Effects of inulin-type fructans on appetite, energy intake, and body weight in children and adults: Systematic review of randomized controlled trials. *Annals of Nutrition and Metabolism, 63*, 42–54.
24. Cani, P., Lecourt, E., Dewulf, E., Sohet, F., Pachikian, B., Naslain, D., et al., (2009). Gut microbiota fermentation of prebiotics increases satietogenic and incretin gut peptide production with consequences for appetite sensation and glucose response after a meal. *American Journal of Clinical Nutrition, 90*, 1236–1243.
25. Cani, P., Joly, E., Horsmans, Y., & Delzenne, N., (2005). Oligofructose promotes satiety in healthy human: a pilot study. *European Journal of Clinical Nutrition, 60*, 567–572.
26. Parnell, J., & Reimer, R., (2009). Weight loss during oligofructose supplementation is associated with decreased ghrelin and increased peptide YY in overweight and obese adults. *The American Journal of Clinical Nutrition, 89*, 1751–1759.
27. Fotiadis, C., (2008). Role of probiotics, prebiotics and synbiotics in chemoprevention for colorectal cancer. *World Journal of Gastroenterology, 14*, 6453–7.
28. Challa, A., (1997). Bifidobacterium longum and lactulose suppress azoxymethane-induced colonic aberrant crypt foci in rats. *Carcinogenesis., 18*, 517–521.
29. Pryde, S., (2002). The microbiology of butyrate formation in the human colon. *FEMS Microbiology Letters, 217*, 133–139.

30. Caderni, G., Femia, A., Giannini, A., Favuzza, A., Luceri, C., Salvadori, M., et al., (2003). Identification of mucin-depleted foci in the unsectioned colon of azoxymethane-treated rats: correlation with carcinogenesis. *Cancer Research, 63*, 2388–2392
31. Lee, J., Shin, J., Kim, E., Kang, H., Yim, I., Kim, J., Joo, H., & Woo, H., (2004). Immunomodulatory and antitumor effects in vivo by the cytoplasmic fraction of Lactobacillus casei and Bifdobacterium longum. *Journal of Veterinary Science, 5*, 41–48.
32. Ghoneum, M., Gallapudi, S., Jason Hamilton, P., & Brown, J. J., (2004). Human squamous cell carcinoma of the tongue exhibits phagocytosis against baker's yeast. *Otolaryngology-Head and Neck Surgery, 131*, 180–180.
33. Takagi, A., Ikemura, H., Matsuzaki, T., Sato, M., Nomoto, K., Morotomi, M., & Yokokura, T., (2008). Relationship between the in vitro response of dendritic cells to Lactobacillus and prevention of tumorigenesis in the mouse. *Journal of Gastroenterology, 43*, 661–669.
34. Serban, D., (2014). Gastrointestinal cancers: Influence of gut microbiota, probiotics and prebiotics. *Cancer Letters, 345*, 258–270.
35. Uccello, M., Malaguarnera, G., Basile, F., D'agata, V., Malaguarnera, M., Bertino, G., et al., (2012). Potential role of probiotics on colorectal cancer prevention. *BMC Surgery, 12*, 35.
36. McFarland, L. V., (2007). Meta-analysis of probiotics for the prevention of traveler's diarrhea. *Travel Medicine and Infectious Disease*, 97–105.
37. Sebastian Domingo, J. J., (2017). Review of the role of probiotics in gastrointestinal diseases in adults. *Gastroenterologia y Hepatologia, 40*, 417–429.
38. Mukai, T., Asasaka, T., Sato, E., Mori, K., Matsumoto, M., & Ohori, H., (2002). Inhibition of binding of helicobacter pylori to the glycolipid receptors by probiotic Lactobacillus reuteri. *FEMS Immunology and Medical Microbiology, 32*, 105–110.
39. Tong, J. L., Ran, Z. H., Shen, J., Zhang, C. X., & Xiao, S. D., (2007). Meta‐analysis: The effect of supplementation with probiotics on eradication rates and adverse events during Helicobacter pylori eradication therapy. *Alimentary Pharmacology and Therapeutics, 25*, 155–168.
40. Rodríguez, J., Martín, M. J., Ruiz, M. A., & Clares, B., (2016). Current encapsulation strategies for bioactive oils: From alimentary to pharmaceutical perspectives. *Food Research International, 83*, 41–59.
41. Zuidam, N. J., & Shimoni, E., (2010). Overview of microencapsulates for use in food products or processes and methods to make them. In: *Encapsulation Technologies for Active Food Ingredients and Food Processing* (pp. 3–29). Springer, New York.
42. Sathyabama, S., Ranjith kumar, M., Bruntha devi, P., Vijayabharathi, R., & Brindha, P. V., (2014). Co-encapsulation of probiotics with prebiotics on alginate matrix and its effect on viability in simulated gastric environment. *LWT – Food Science and Technology, 57*, 419–425.
43. Onwulata, C. I., (2012). Encapsulation of new active ingredients. *Annual Review of Food Science and Technology, 3*, 183–202.
44. Abd El-Salam, M. H., & El-Shibiny, S., (2015). Preparation and properties of milk proteins-based encapsulated probiotics: A review. *Dairy Science and Technology, 95*, 393–412.

45. Zhao, R. X., Sun, J. L., Torley, P., Wang, D. H., & Niu, S. Y., (2008). Measurement of particle diameter of Lactobacillus acidophilus microcapsule by spray drying and analysis on its microstructure. *World Journal of Microbiology and Biotechnology, 24*, 1349–1354.
46. Vieira da Silva, B., Barreira, J. C. M., & Oliveira, M. B. P. P., (2016). Natural phytochemicals and probiotics as bioactive ingredients for functional foods: Extraction, biochemistry and protected-delivery technologies. *Trends in Food Science and Technology, 50*, 144–158.
47. Santiago, L. G., & Castro, G. R., (2016). Novel technologies for the encapsulation of bioactive food compounds. *Current Opinion in Food Science, 7*, 78–85.
48. Bigliardi, B., & Galatib, F., (2013). Innovation trends in the food industry: The case of functional foods. *Trends in Food Science and Technology, 31*, 118–129.
49. De Prisco, A., & Mauriello, G., (2016). Probiotication of foods: A focus on microencapsulation tool. *Trends in Food Science and Technology, 48*, 27–39.
50. Moumita, S., Goderska, K., Johnson, E. M., Das, B., Indira, D., Yadav, R., Kumari, S., et al., (2017). Evaluation of the viability of free and encapsulated lactic acid bacteria using in-vitro gastrointestinal model and survivability studies of symbiotic microcapsules in dry food matrix during storage. *LWT-Food Science and Technology, 77*, 460–467.
51. Pauck, C., De Beer, D., Aucamp, M., Liebenberg, W., Stieger, N., Human, C., et al., (2017). Inulin suitable as reduced-kilojoule carrier for production of microencapsulated spray-dried green *Cyclopia subternata* (honeybush) extract. *LWT Food Science and Technology, 75*, 631–639.
52. Krasaekoopt, W., & Watcharapoka, S., (2014). Effect of addition of inulin and galactooligosaccharides on the survivals of microencapsulated probiotics in alginate beads coated with chitosan in simulated digestive system, yogurt and fruit juice. *LWT Food Science and Technology, 57*, 761–766.
53. Amakiri, A. C., Kalombo, L., Thantsha, M. S., Bigliardi, B., & Galatib, F., (2015). Lyophilized vegetal BM 297 ATO-Inulin lipid based synbiotic microparticles containing Bifid bacterium longum LMG 13197: design and characterisation. *Journal of Microencapsulation*, 1–8.
54. Phoem, A. N., Chanthachum, S., & Voravuthikunchai, S. P., (2015). Applications of microencapsulated bifid bacterium longum with eleutherine Americana in fresh milk tofu and pineapple juice. *Nutrients, 7*, 2469–2484.
55. García, P., Vega, J., Jimenez, P., Santos, J., & Robert, P., (2013). Alpha-tocopherol microspheres with cross-linked and acetylated inulin and their release profile in a hydrophilic model. *European Journal of Lipid Science and Technology, 115*, 811–819.
56. Darjani, P., Nezhad, M. H., Kadkhodaee, R., & Milani, E., (2016). Influence of prebiotic and coating materials on morphology and survival of a probiotic strain of Lactobacillus casei exposed to simulated gastrointestinal conditions. *LWT Food Science and Technology, 73*, 162–167.
57. Sultana, K., Godward, G., Reynolds, N., Arumugaswamy, R., Peiris, P., & Kailasapathya, K., (2000). Encapsulation of probiotic bacteria with alginate–starch and evaluation of survival in simulated gastrointestinal conditions and in yoghurt. *International Journal of Food Microbiology., 94*, 323–328.

58. Mitropoulou, G., Nedovic, V., Goyal, A., & Kourkoutas, Y., (2013). Immobilization Technologies in Probiotic Food Production. *Journal of Nutrition and Metabolism, 13*, 1–15.
59. Kumar, B. V., Vijayendra, S. V. N., & Reddy, O. V. S., (2015). Trends in dairy and non-dairy probiotic products. *Journal of Food Science and Technology, 52*, 6112–6124.
60. Yucel Falco, C., Sotres, J., Rascón, A., Risbo, J., & Cárdenas, M., (2017). Design of a potentially prebiotic and responsive encapsulation material for probiotic bacteria based on chitosan and sulfated β-glucan. *Journal of Colloid and Interface Science, 487*, 97–106.
61. Dos Santos, C., Buera, P., & Mazzobre, F., (2017). 'Novel trends in cyclodextrins encapsulation. Applications in food science.' *Current Opinion in Food Science, 16*, 106–113.
62. Tampau, A., González-Martinez, C., & Chiralt, A., (2017). Carvacrol encapsulation in starch or PCL based matrices by electrospinning. *Journal of Food Engineering, 214*, 245–256.
63. Ghorani, B., & Tucker, N., (2015). Fundamentals of electrospinning as a novel delivery vehicle for bioactive compounds in food nanotechnology. *Food Hydrocolloids, 51*, 227–240.
64. Dias, D. R., Botrel, D. A., Fernandes, R. V. D. B., & Borges, S. V., (2017). Encapsulation as a tool for bioprocessing of functional foods. *Current Opinion in Food Science, 13*, 31–37.
65. Burgain, J., Gaiani, C., Linder, M., & Scher, J., (2011). Encapsulation of probiotic living cells: from laboratory scale to industrial applications. *Journal of Food Engineering, 104*, 467–483.
66. De Preter, V., Hamer, H., Windey, K., & Verbeke, K., (2011). The impact of pre- and/or probiotics on human colonic metabolism: Does it affect human health?. *Molecular Nutrition and Food Research, 55*, 46–57.
67. Iravani, S., Korbekandi, H., & Mirmohammadi, S. V., (2015). Technology and potential applications of probiotic encapsulation in fermented milk products. *Journal of Food Science and Technology, 52*, 4679–4696.
68. Pagnini, C., Saeed, R., Bamias, G., Arseneau, K., Pizarro, T., & Cominelli, F., (2010). Probiotics promote gut health through stimulation of epithelial innate immunity. *Proceedings of the National Academy of Sciences, 107*, 454–459.
69. Valcheva, R., Koleva, P., Meijer, B. J., Walter, J., Gänzle, M., Dieleman, L. A. (2012). 1091a Beta-Fructans Reduce Inflammation in Mild to Moderate Ulcerative Colitis Through Specific Microbiota Changes Associated With Improved Butyrate Formation and MUC2 Expression. *Gastroenterology, 142*, 196.
70. Pinto, S., Fritzen-Freire, C. B., Muñoz, I., & Amboni, R. (2012). Effects of the addition of microencapsulated Bifidobacterium BB-12 on the properties of frozen yogurt. *Journal of Food Engineering. 111*(4), 563–569.

CHAPTER 3

CAROTENOID COMPOUNDS: PROPERTIES, PRODUCTION, AND APPLICATIONS

VICTOR NAVARRO-MACÍAS,[1] AYERIM HERNÁNDEZ-ALMANZA,[1] JANETH VENTURA,[2] MÓNICA L. CHÁVEZ-GONZÁLEZ,[1] DANIEL BOONE-VILLA,[3] JOSÉ LUIS MARTÍNEZ,[1] JULIO CESAR MONTAÑEZ,[1] and CRISTÓBAL N. AGUILAR[1]

[1] *Food Research Department, School of Chemistry, Autonomous University of Coahuila, Saltillo, 25280, Coahuila, E-mail: cristobal.aguilar@uadec.edu.mx*

[2] *School of Health Sciences, University of Coahuila, Piedras Negras, Coahuila 26090, México*

[3] *School of Medicine Unit North, University of Coahuila, Piedras Negras, Coahuila, 26090, México*

ABSTRACT

Carotenoids are compounds of lipophilic nature, consisting of 40 carbon molecules. Carotenoids are important compounds in the cosmetic alimentary and pharmaceutical industry for its various applications such as colorants. The demand for carotenoids is increasing not only because they are natural colorants, but also for its antioxidant activity that helps to prevent cardiovascular diseases, age-related macular degeneration, multiple sclerosis and some types of cancer. Carotenoids are found in a large number of natural products, such as fruits, vegetables, and seafood. Although the highest production of carotenoids is obtained from plants, several studies show that an alternative to satisfy the demand in the market, is the production of carotenoids from microorganisms.

3.1 INTRODUCTION

Several epidemiological studies have shown that consumption of fruits and vegetables in daily diet could help to prevent no communicable diseases such as cardiovascular disease and some cancer types. In such a way that the mortality rate could be reduced, upto 1.7 million lives could be saved if adequate intake existed [1]. Fruits and vegetables are important sources of vitamins (A, B, C, and E) and minerals such as calcium, phosphorus, and magnesium, in addition to a wide variety of bioactive compounds such as polyphenols and carotenoids [2].

Carotenoids are liposoluble pigments that are distributed among living beings. Currently, more than 700 carotenoids have been characterized in nature [3]. These compounds are responsible for giving color to some fruits. For example, green vegetables contain high amounts of hydrocarbon carotenes and xanthophylls; lycopene is a red lipophilic pigment present in mature tomatoes; orange color of carrots is caused by β-carotene; capsanthin is responsible for the bright red pigment of peppers; the pink/red coloration of crustaceans is due to astaxanthin, among others [4, 5]. These pigments are responsible for the wide variety of orange-red colors seen in nature that absorbs light in the wavelength range of 300–600 nm. Absorbance is directly related to the number of conjugated double bonds and functional groups present in the structure.

In addition to coloring, carotenoids are natural substances with important biological activities, including antioxidant properties and some of them are precursors of vitamin A (β-carotene, γ-carotene and β-cryptoxanthin) [6]. Due they protect cells from oxidative damage they can prevent chronic diseases such as cancer, cardiovascular disease, diabetes, among others [7]. Molecular structure of carotenoids will define the chemical properties and these will in turn determine the physical properties and biological functions of these compounds [8, 9]. In this way, carotenoids have applications in the cosmetic, pharmaceutical and medical industries [10].

Commercially, carotenoids are obtained from natural products [11], however, with the increase in demand in the market (annual rate of 2.9%) have sought alternatives for their production, one of them is the biotechnological production through microorganisms [12–14]. However, there are still some challenges such as reducing production costs and increasing production. These compounds are found in plants in greater concentration and variety, but they can also be synthesized by some microorganisms such as bacteria, yeasts and filamentous fungi such as *Sporobolomyces, Xanthophyllomyces, Rhodosporidium, Blakeslea trispora, Rhodotorula, Cryptococus,* and *Sporidiobolus* [15, 16].

3.2 CHEMISTRY OF CAROTENOIDS

Carotenoids are synthesized by the binding of two geranyl-geranyl diphosphate C–20 molecules. All carotenoids contain a poly-isoprenoid structure, a long double bond conjugate chain and an almost bilateral symmetry around the central double bond. Carotenoids can be open-chain or have cycles at one or both ends of the chain, circulation and other modifications such as isomerization or migration of double bonds, will result in various structures. For this reason, carotenoids show a great variety of spectroscopic and functional properties [17]. They are classified into carotenes, hydrocarbon compounds, which can be divided into provitamin A (β-carotene and α-carotene) and non-provitamin A (lycopene and torulene) [5], xanthophylls which in addition to having carbon and hydrogen have oxygen in their structure, which may be as a group: hydroxyl, methoxy, epoxy, carboxyl or carbonyl (Figure 3.1) [18–20].

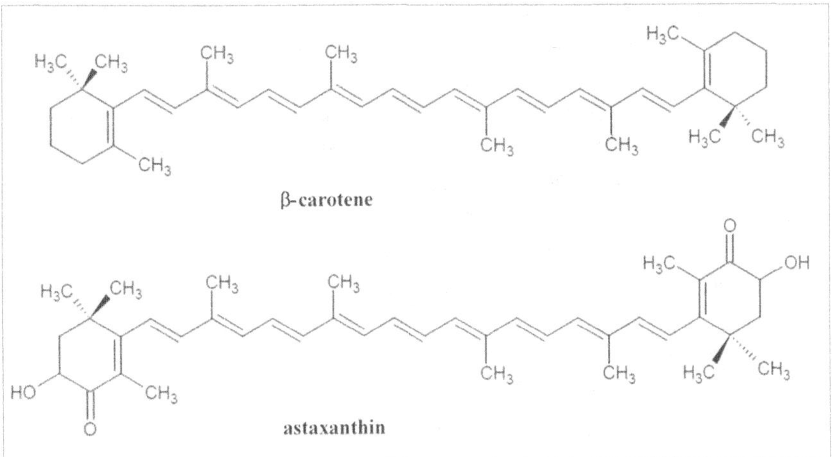

FIGURE 3.1 Chemical structure of carotenoids.

Due to the presence of conjugated double bonds, the carotenoids can undergo isomerization to cis-trans isomers, factors such as temperature, light and pH can also produce alterations in carotenoids (photochemical reactions oxygenation of the double bonds to epoxide, hydroxyl and peroxides functions), as well as its nutritional value [19–22]. Carotenoids have been shown to exhibit greater stability when in their natural matrix. The oxidation process of carotenoids is more frequent when the cell wall has been affected by some process of rupture, for this reason the carotenoids lose stability once they have been extracted in oils or solvents [18, 23].

3.3 APPLICATION OF CAROTENOIDS

Due to the structural variety and the various biologic functions such as species-specific coloring, photo protection and light harvesting, as well as serving as precursors for many hormones [24]. In the food industry, they have been used as additives and dyes in some foods and beverages such as orange juice, lemon, nectars, butter and sausage. In addition to coloring, carotenoids play an important role in human health because of their beneficial effects, currently carotenoids have application in the pharmaceutical and cosmetic industry [25].

The beneficial properties of carotenoids are associated with various factors (chemical structure, including the number of conjugated double bonds, type of structural end groups and oxygen-containing substituents) and the position it occupies in the lipid membrane (Figure 3.2) [26–28]. Some of the most commonly used carotenoids are β-carotene, astaxanthin, canthaxanthin, lycopene, zeaxanthin and lutein. In the case of canthaxanthin and astaxanthin have been used for tissue pigmentation of salmon, trout and crustaceans mainly, in order to achieve a desirable color to meet market demand. There are reports that carotenoids in addition to coloration fulfill some biological functions in fish such as an antioxidant defense, immune function and vitamin precursors [25, 29, 30]. Astaxanthin is also used in the food industry as a supplement for animal feed, pharmaceuticals and cosmetics as a colorant and antioxidant [25]. It has been shown that astaxanthin has an antioxidant capacity 500 times higher than vitamin E and 10 times more than β-carotene [31]. Reports indicate that astaxanthin may also have health benefits in areas such as eye and brain health, UV protection, and cardiovascular health among others [32].

Carotenoid Compounds: Properties, Production, and Applications 67

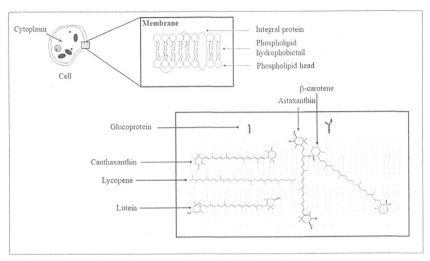

FIGURE 3.2 Position of the carotenoids in the lipid membrane.

Carotene is another biologically active molecule, which are related to its antioxidant power in the inactivation of certain species of reactive oxygen, such as singlet oxygen [33]. It has been mentioned that carotenoids have antioxidant functions, and may reduce the risk of type 2 diabetes by reducing oxidative stress, which plays an important role in the development of diabetes. It was reported that diets high in β-carotene and α-carotene are associated with the reduction of type 2 diabetes in generally healthy men and women. Studies have also shown that β-carotene can reduce tumorigenesis by up to 60% in neuroblastoma in vivo inducing neuronal differentiation [34]. β-carotene and α-carotene are also precursors of vitamin A. Vitamin A is an important micronutrient to prevent blindness, helps immune function, growth, development and gastrointestinal functioning of the body [35]. Due to this, the demand of this carotenoid as additive in cosmetics and application in foods has been increasing [33]. Other carotenoids such as lutein and zeaxanthin do not have provitamin A capacity [36], however, these xanthophylls have been associated with the prevention of ocular diseases. In the case of lutein, there are reports that it reduces the risk of eye-related diseases such as cataracts and macular degeneration (AMD) [37, 38]. Lutein and zeaxanthin can also reduce light-induced skin damage, lutein has been reported in epidemiological studies to reduce the risk of some chronic diseases such as colon cancer, breast cancer, type 2 diabetes, and cardiac disease [39, 40]. The protective role of lutein and zeaxanthin in both, retina and the lens, is due to its ability to filter harmful damaging short-wave blue light to function as antioxidants and to stabilize the integrity of the membrane [36, 41, 42]. It is believed that these biological functions play an important role in helping to reduce the oxidative damage induced by light from reactive oxygen species [43, 44]. These carotenoids are marketed as dietary supplements [45] since lutein and zeaxanthin like other carotenoids are not produced by the human body and must be consumed in the diet [42]. However, little or no water solubility causes the carotenoids to lack the desired bioavailability [46], while carotenoids from dietary supplements are better absorbed due to the oil content present in the supplement.

Lycopene is a red pigment has applications in the cosmetic industry as a dye and as an additive in the food industry. Lycopene is a carotene with high antioxidant power, with a singlet oxygen extinction capacity twice as high as that of β-carotene and 10 times higher than that of α-tocopherol [47, 48]. Several reports indicate that low concentrations of lycopene help to reduce reactive oxygen species as mentioned. Campos et al. [49] they report a decrease in cell cytotoxicity and the production of reactive oxygen species in cells cultured in vitro, while in vivo they observed several beneficial effects of lycopene on redox imbalance and inflammation in animals that had been exposed to smoke cigarette for a short time (5 days). On the other hand Sahin et al. [50] mention that dietary supplementation with lycopene reduces the number and size of renal carcinomas that develop in rats, this could be generalized for the humans. Lycopene may also be useful in preventing gastric damage, the results of Boyacioglu et al. [51] show that 100 mg/kg of lycopene significantly inhibited indomethacin-induced gastric damage. Unlike β-carotene and lycopene has no provitamin A activity due to the absence of the γ-ionic ring common in this carotenoid [52]. The literature reports mention that the concentration of lycopene in blood is inversely related to the incidence of cardiovascular diseases and some cancers such as breast, cervical, ovary, endometrial, lung, bladder, oral cavity, esophagus, stomach, colon, pancreas, and prostate [50, 53].

3.4 SOURCES OF CAROTENOIDS

In developed countries, 80–90% of carotenoid intake comes from fruit and vegetable consumption [4, 21]. Of the more than 700 natural carotenoids identified so far [54], up to 40 are present in the human diet and can be absorbed and metabolized by the human body. However, only about 20 have been identified, of which only six (β-carotene, β-cryptoxanthin, α-carotene, lycopene, lutein and zeaxanthin) represent most of the total carotenoids present in blood and tissues humans. These carotenoids have been studied and associated with some health benefits [21].

Carotenoids are natural pigments that go from yellow to red through orange hues, which color fruits, vegetables, leaves and flowers (Table 3.1) [18, 55, 56]. They are synthesized by plants (Figure 3.3). In vegetables is where the greatest variety and concentration of carotenoids is found, however, we can find them in bacteria, algae and fungi [11, 57, 58], while animals get carotenoids from their diet [59]. In plants, algae and phototrophic bacteria, carotenoids (mainly xanthophylls) are localized to lipid membranes or stored in plasma vacuoles (specific pigment-protein complexes) [60]. Carotenoids are essential for the capture of energy from the solar emission spectrum and constitute the basic structural units of the photosynthetic apparatus. Of the many natural carotenoids, less than 50 play a light-gathering role in photosynthetic organisms such as lutein and β-carotene [61–63].

In addition to light harvesting, carotenoids also act as photoprotectors by cooling excess energy under high-light stress and preventing the formation of highly reactive singlet oxygen, which has deleterious and potentially lethal effects [64, 65]. Plant carotenoids also play a vital ecological role, which is important for the survival and propagation of species [66]. One of the factors that influence the presence of carotenoids are genotype, there is an alternative set of biosynthetic genes of carotenoids that exert their effects during the ripening of the fruit, some genes show enhanced maturation expression patterns such as Ggpps–2, Psy–1, CRTISO, CYC-B y CrtR-b1. Other factors are pre-harvest handling, maturity, during this process cellular changes occur, one of them is the transition of chloroplasts present in the immature fruit to chromoplasts present in the fruit in mature stage, and this event is accompanied by a qualitative change in the carotenoid profile. During storage and processing operations carotenoids are affected by oxidation and the structural changes which are common

TABLE 3.1 Distribution of Carotenoids in Various Foods

Food	Carotenoids	Reference
Carrot	α-caroteno and β-carotene	[1]
Tomato	Lycopene	[2]
Watermelon	Lycopene	[3]
Orange	Violaxantina, β-cryptoxanthin, luteina, zeaxantina	[4]
Papaya	Lycopene	[5]
Mango	α-carotene, β-carotene, and β-cryptoxanthin	[6]
Spinach	Luteín, zeaxanthin and β-carotene	[7]
Maize	Luteín and zeaxanthin	[6]
Food	Major Carotenoids	
Carrot (*Daucus carota*)	α and β-carotene	
Orange (*Citrus sinensis*)	Violaxantina, β-cryptoxanthin, luteina, zeaxantina	
Mango (*Mangifera indica*)	Violaxanthin, β-carotene, Lyopene	
Tomato (*Lycopersicum esculentum*)	Lycopene	
Red bell pepper (*Capsicum anuum*)	Capsanthin, capsorrubina	
Melocoton (*Prunus persica*)	β-cryptoxanthin, lutein	
Papaya (*Carica papaya*)	β-cryptoxanthin, β-carotene	
Guava (*Psidium guajava*)	Lycopene, β-carotene	
Plum (*Spondias lutea*)	β-cryptoxanthin	
Maize (*Zea mays*)	Luteín and zeaxanthin	
Spinach (*Spinacia oleracea*)	Luteín and zeaxanthin	

when applying heat. Packaging at low temperatures and in oxygen-free atmospheres can help maintain the carotenoid content in food [56, 67, 68].

3.5 MICROBIAL PRODUCTION

Carotenoids have taken great interest due to high demand in the pharmaceutical, cosmetic and food industries [69]. In order to satisfy this demand, alternative natural sources have been sought for the production of these compounds [70–72]. Although few carotenoids are commercially produced, the highest production of carotenoids is obtained from natural sources such as tomatoes or carrots [73]. However, the fermentative route

Carotenoid Compounds: Properties, Production, and Applications

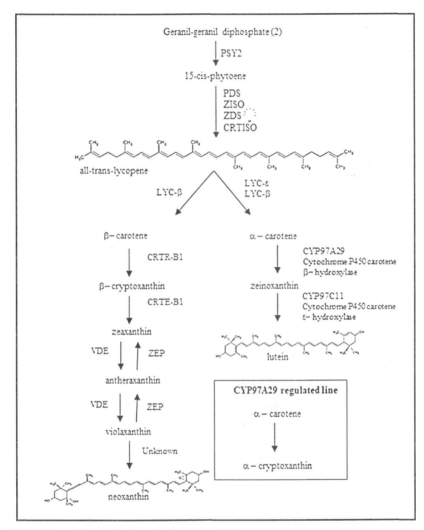

FIGURE 3.3 Biosynthetic route of carotenoids in tomato.

has become a viable option for the synthesis of carotenoids. Carotenoids, produced by microorganisms form an important class of natural pigments for biotechnological applications. As in plants and animals, it is proposed that pigments play important physiological roles in microbial cells, mainly those associated with the response to stress [8, 24, 74, 75].

In the case of some microorganisms, production of carotenoids occurs in the exponential and continuous phase during the stationary phase and the presence of a suitable carbon source it is very important for the biosynthesis of carotenoids for the phase of cell death [76]. However, the high costs of carotenoid production may limit the use of microorganisms so an alternative is through the use of agro-industrial waste as a substrate [77]. In addition, reducing costs can minimize the environmental and energy problems related to its elimination [76]. The research group of Buzzini and Martini from the Dipartamento di Biologia Vegetale of Universita di Perugia, Italy, reports the use of grape must, glucose syrup, sugar beet molasses, soy flour extract and cornflour extract, obtaining the best yield with grape must as the only source of carbon (5.95 mg/L of total carotenoids and 630 µg/g cells in dry weight), they also report that the accumulation of torularodine is greater than those of β-carotene and torulene. On the other hand, Aksu and Eren from the Department of Chemical Engineering of Hacettepe University, Turkey evaluated sucrose, molasses and lactose serum, obtaining the maximum concentration of carotenoids (125 mg/L) using 20 g/L of sucrose molasses, however, the maximum yield (35.5 mg of total carotenoids per gram of dry cells) was achieved when 13.2 g/L lactose serum was added to the culture medium. Other authors report the use of fermented corn liquor, boiled rice water and crude glycerol [14, 78, 79]. In addition to the substrate, fermentation conditions also influence the type and concentration of carotenoids [77], such as light.

Light is an environmental factor that helps regulate the metabolism of microorganisms, in the case of fungi it has been reported that several physiological processes are influenced by light [80]. However, the intensity and protocol of illumination vary with the microorganism. In the production of carotenoids, light plays an important role since it can activate enzymes that participate in the biosynthesis of carotenoids [3]. Some functions of carotenoids are associated with their ability to absorb light. These compounds participate in the deactivation of free radicals that occur during the normal metabolism of the cells. It has been shown that some microorganisms inactivate $^1O_2$, OH radicals, and other oxidants by transferring energy from high levels of excitation to a carotenoid triplet. One biological sources of $^1O_2$ production include energy transfer from excited photosensitizers, from the enzymatic reactions (Figure 3.4) [80, 81]. A study by Zhang et al. [82] showed that light affects carotenoid production by *Rhodotorula glutinis*, the concentrations obtained were (2.6 mg/L) with light and (1.2 mg/L) without light. The results obtained by these authors, in addition to showing an increase in pigment concentration, reduced production time. Similar results are shown by Khanafari et al. [80] when evaluating different types of light found that both light blue and white light promoted the production of carotenoids by *Mucor hiemalis* (1.2 mg/g) and (1.3 mg/g), respectively, three times more than production in the absence of light (0.45 mg/g). Several reports indicate that the stress caused by light is similar to that caused by the substrate and temperature.

Temperature is a parameter that affects some enzymes involved in microbial growth and carotenoid biosynthesis [3, 77]. The effect of temperature varies according to the microorganism, this effect is seen in the amount of synthesized carotenoids. In some yeasts such as *Rhodotorula mucilaginosa*, *Phaffia rhodozyma* and *Xanthophyllomyces dendrorhous*, temperatures above 30°C reduce the synthesis of carotenoids and less than 20°C inhibit microbial growth and synthesis of carotenoids, a study shows that growth and the carotenoid formation rates of the yeast increased with increasing temperature to 30°C and decreased sharply above 30°C. This was due to the denaturation of the enzyme system of the microorganism at higher temperatures. The temperature also influences the type

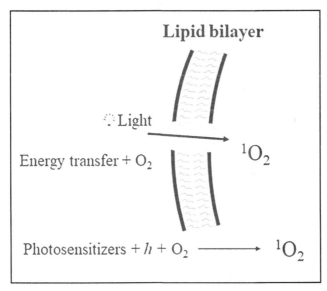

FIGURE 3.4 Sources of singlet oxygen.

of carotenoids that are produced as demonstrated by Cheng and Yang [83] when evaluating temperatures from 20°C to 30°C. In this research work was obtained the optimum temperature for the highest concentration of carotenoids (25°C). Concentration of carotenoids was 2506 µg/L. They also report that at high temperatures the accumulation of β-carotene and torulene decreases while the concentration of torularhodin increases.

pH is another important factor in carotenoid production as temperature or light, pH directly affects microbial growth as well as carotenoid production, as demonstrated by Bhosale and Gadre [84] when evaluating the effect of pH on strains of *R. glutinis* and observed that at extreme pH values there was little growth and less carotenoid production although the proportion of carotenoids was not altered. On the other hand, Aksu and Eren [77] reported that the production of carotenoids and the production of biomass increase as the pH increases, however, when it exceeds a pH of 6, the growth and production of carotenoids is affected. The best yield was given at pH 6 obtaining 13.5 mg/mL. These results are not similar to those obtained by Cheng and Yang [83]. These authors evaluated four pH values (4, 5, 6, and 7) and reported that the maximum concentration (317.6 µg/g) of pigments was obtained at pH 5. They also observed that as the pH increases, the accumulation of torulene and torularhodine increases while that of β-carotene decreases. The effect of the factors mentioned above will affect differently according to the microorganism used for the production of carotenoids.

3.6 METHODS OF CAROTENOID RECOVERY

Carotenoids are usually stable within a cell but once the carotenoids are released, various factors such as light, temperature, pH and oxygen, can influence the stability of these molecules, causing structural changes which may affect biological activity or its beneficial properties for health [18–20, 80, 85]. There are several methods for pigment recovery. Traditional methods such as maceration are too late and in some cases, require high amounts of solvents. However, to date several methods have been developed for the extraction of pigments. Some methodologies that have been evaluated for the extraction of carotenoids are ultrasound, microwaves, supercritical fluids, steam explosion, enzymatic lysis, electric pulses among others [86–88].

The solid-liquid extraction of the target compound from the plant materials is a unitary operation consisting of the separation of the solute (soluble compounds) from the plant matrix by the use of a liquid solvent [89, 90]. The extraction efficiency depends on the nature of the plant matrix and the solute to be extracted, from the operating conditions of the extraction process such as pressure and temperature which modify the physical properties of the solvent to reduce the surface tension of the solvent. In these methods the extraction efficiency of carotenoids can be improved by using combinations of solvents to facilitate partitioning. Two solvent methods are: low-pressure solvent extraction, including Soxhlet extraction. This method provides the transfer of soluble components (oil) from an inert material (grease matrix) to a solvent (hexane or petroleum ether) whose matrix is in contact [91]. Another method with solvents is liquid extraction under pressure. It is a technology that uses liquid solvents such as hexane, ethanol and acetone to recover objective compounds in a shorter extraction time compared to the Soxhlet process. Pressurized liquids have the advantage of a higher solubility with an increase in temperature due to a greater diffusion of the analyte from the solid matrix to the bulk solvent and the reduction of the viscosity of the solvent, which facilitates the penetration of the solvent into the matrix [89, 92–94]. The research group of Kwang et al. [92] compared these two methods of extraction and have found that extraction with pressurized liquids had a better yield (0.5 mg/g sample) than extraction with Soxhlet (0.26 mg/g sample) This result is mainly due to the temperature and high pressure that exists in the extraction process. On the other hand, Cardenas-Toro et al. [89], when comparing the aforementioned methods, conclude that extraction with pressurized liquids has better yields than extraction by Soxhlet, obtaining as results using liquids under pressure (305 ± 18 mg α-carotene/g extract and 713 ± 46 mg β-carotene/g extract) exceeding Soxhlet extraction (142 ± 13 mg a-carotene/g extract and 317 ± 46 mg b-carotene/g extract). A disadvantage of the use of organic solvents is that it may limit the use of carotenoids, for which reason alternative methods have been sought for the recovery of these pigments.

Although the extraction of carotenoids are carried out mainly with solvents there are other methods that have been reported in the literature such as supercritical fluids. This technique may be an alternative for the recovery of carotenoids [95]. Mainly CO_2 is used as the extractant and a gentle extraction of the biologically active labile substances is provided. Low temperature is an important factor in the extraction of thermolabile substances such as carotenoids [85, 86, 90, 96]. Supercritical carbon dioxide has been reported to be a suitable solvent for the extraction of carotenoids because of the low polarity of the compounds. The best extraction yields of carotenoids, using *D. salina* as raw material, are obtained at the maximum operating temperature (60°C) and at a pressure of approximately 400 bar [97].

Assisted microwave extraction involves the use of microwave energy to heat a solvent in contact with a sample to allow the release of a bioactive compound from the sample matrix into the solvent. Hiranvarachat et al. [23] compare microwave extraction with the Soxhlet method, the obtained results showed that a greater extraction was obtained in the method of Soxhlet (61.13 ± 3.6 mg/g d.b.), however, the extraction time was 6 h, although the concentration extracted with the microwave was lower (51.79 ± 2.11 mg/g d.b) and the extraction time was 3 min. A further study by Hiranvarachat and Devahastin, [98] evaluated the effect of intermittent microwave radiation on low power (180 W) and high power (300 W), as well as the amount of solvent resulting in a concentration of 58 ± 6 Mg/100 g (d.b.), of β-carotene. On the other hand, Ho et al. [99] report that optimal conditions for the extraction of lycopene from tomato is a ratio of solvent 0:10 to 400 W with a yield of 13,592 mg/100 g of lycopene and that the ethyl acetate was a better solvent for the recovery of lycopene compared to hexane.

Ultrasound is another method widely studied. Compared to the extraction with microwaves and with supercritical fluids, ultrasound is less expensive, low energy requirements and low consumption of solvents [88, 97, 100]. Extraction efficiency using ultrasound is mainly attributed to the effect of acoustic cavitation produced in the solvent as a result of the passage of ultrasonic waves. Ultrasound also exerts a mechanical effect by improving the penetration of the solvent into the tissue, increasing contact over the surface area between the solid and liquid phase. As a result, the solute diffuses rapidly from the solid phase to the solvent [88, 101]. Ultrasound offers a net advantage in terms of productivity, performance and selectivity, with improved processing time, improved quality, reduced chemical and physical risks, and is environmentally friendly [102]. According to Dey and Rathod, [103] the conditions for the use of this method influence the yield, the optimum conditions for the extraction of β-carotene from Spirulina were 1.5 g of Spirulina (previously treated with Methanol) in 50 ml of n-heptane at 30°C, 167 W/cm^2 of electrical acoustic intensity and 61.5% duty cycle for 8 min with a probe tip length of 0.5 cm immersed in the extraction solvent from the surface. The maximum extraction obtained under the optimal parameters mentioned above was 47.10%. Another study by Eh and Teoh, [104] where optimization of lycopene extraction, yielding the average relative lycopene yield of 99% at 45.6 min, 47.6°C and the ratio of the solvent to the lyophilized tomato sample (w/P) of 74.4: 1. From the optimized model, the average yield of lycopene was 5.11 ± 0.27 mg/g dry weight. In both investigations, they conclude that the ultrasound promotes the extraction yield and minimizes the degradation of the carotenoids. The quantification and separation of carotenoids are carried out mainly by chromatographic methods, one of them is high-performance liquid chromatography (HPLC) [105, 106].

The recovery of carotenoids is a very important stage and can be complicated since this depends on the use of carotenoids in the pharmaceutical and cosmetic food industry [23].

3.7 RECENT STUDIES ABOUT PRODUCTION AND RECOVERY OF MICROBIAL PIGMENTS IN THE FOOD RESEARCH DEPARTMENT – U.A.C., MEXICO

In Mexico, exist about 30 public institutions and universities dedicated to biotechnology [107]. Each research center has a different approach in order to meet the needs of the population. Currently, the Food Research Department of Autonomous University of Coahuila develop studies about microbial production of pigments, for example, the molecular identification of pigments-producer strains, various techniques to recovery as well as the evaluation of the stability of microbial pigments under different environmental conditions has been studied (Figure 3.5). *Rhodotorula sp., Penicillium purpurogenum and Chlorella vulgaris* are the principal microorganism employed in the pigment production.

Espinoza-Hernández et al. [108] characterized morphological, physiological and molecularly three fungal strains isolated from *Quercus* sp and *Larrea tridentata*. The strains were identified as *P. purpurogenum* GH2 and *P. pinophilum* EH2 and EH3, and the authors reported that the strain

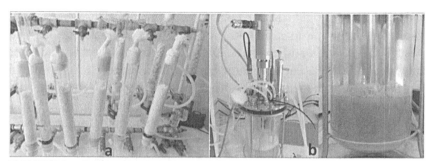

FIGURE 3.5 Carotenoids production. (a) Carotenoids production by *R. glutinis* YB-252 under solid state fermentation using polyurethane foam such as support and glucose such substrate, after 72 h of fermentation. (b) Production of carotenoids by *R. glutinis* P4M422 in a bioreactor using goat milk whey after 72 h of fermentation.

P. purpurogenum GH2 produced the highest amount of pigments under submerged state fermentation, pH 10 and 24°C.

Due this result, *P. purpurogenum* GH2 has been studied to the red pigment production, current studies are focus in the effect of ohmic heating processing conditions on color stability [109], also, the effect of heat exposure on the color intensity was studied [110].

On the other hand, there is special attention in carotenoids production by microorganisms, in this case, the red yeast *Rhodotorula glutinis* has been employed and the different culture conditions such as pH, temperature, substrate, were evaluated, even, the production under solid and submerged state fermentation was studied. Hernández-Almanza et al. [73] evaluated the conditions of solid-state fermentation to produce total carotenoids by *R. glutinis* YB-252. In the study, polyurethane foam was used like an inert support and various factors were evaluated under Plackett-Burman design. The authors report that the highest carotenoids concentration (340 mg/L) was obtained at pH 4, 25°C, moisture 90%, carbon source 40 g/L, during 48 h. Also, in this study, the specific production of lycopene was reported, and the major concentration of lycopene (6.82 mg/L) was obtained with the addition of imidazole 250 ppm. Imidazole and another nitrogen- heterocyclic compounds are effective in the inhibition of enzymes involved in the carotenoid biosynthetic pathway.

R. glutinis P4M422 is a red yeast also studied to carotenoids production. Mata Gómez [111] evaluated the effect of medium composition based on goat milk whey on the carotenoids production. Different carbon and nitrogen source such as glucose, galactose, lactose, starch, ethanol, urea, yeast extract, etc., were reviewed. The author indicates the highest carotenoids concentration (995 µg/L) was observed in the treatment with ethanol at 10% (v/v) and the addition of urea as sole nitrogen source (3018 µg/L of total carotenoids).

3.8 PROSPECTS

Carotenoids have been extensively studied and have been shown to have beneficial health properties, but there are still difficulties in the process of obtaining them: (a) microbial production, despite being a good alternative has not yet a standardized process to meet the high demand for these natural compounds, production costs are still high and as a consequence the product has a high cost; (b) the low stability of carotenoids outside their matrix and the use of solvents limits the use of these compounds in the food and pharmaceutical industry as environmental factors can structurally alter these compounds, altering their benefits.

KEYWORDS

- cardiovascular diseases
- microorganisms
- nutritional supplement

REFERENCES

1. Mustafa, A., Trevino, L. M., & Turner, C., (2012). "Pressurized hot ethanol extraction of carotenoids from carrot by-products," *Molecules*, *17*(2), 1809–1818.
2. Martí, R., Roselló, S., & Cebolla-Cornejo, J., (2016). "Tomato as a source of carotenoids and polyphenols targeted to cancer prevention," *Cancers*, *8*(6), 1–28.
3. Oberoi, D. P. S., & Sogi, D. S., (2017). "Utilization of watermelon pulp for lycopene extraction by response surface methodology," *Food Chemistry*, *232*, 316–321.
4. Cazzonelli, C. I., (2011). "Carotenoids in nature: Insights from plants and beyond," *Functional Plant Biology*, *38*(11), 833–847.

5. Krinsky, N. I., & Johnson, E. J., (2005). "Carotenoid actions and their relation to health and disease," *Molecular Aspects of Medicine, 26*(6), 459–516.
6. Zhou, C. H., Xu, C. J., De Sun, C., Li, X., & Chen, K. S., (2007). "Carotenoids in white- and red-fleshed loquat fruits," *Journal of Agricultural and Food Chemistry, 55*(19), 7822–7830.
7. Saini, R. K., Nile, S., H., & Park, S. W., (2015). "Carotenoids from fruits and vegetables: Chemistry, analysis, occurrence, bioavailability and biological activities," *Food Research International, 76,* 735–750.
8. Marova, I., Carnecka, M., Halienova, A., Certik, M., Dvorakova, T., & Haronikova, A., (2012). "Use of several waste substrates for carotenoid-rich yeast biomass production," *Journal of Environmental Management, 95,* 338–342.
9. Rodrigues, E., Mariutti, L. R. B., Chisté, R. C., & Mercadante, A. Z., (2012). "Development of a novel micro-assay for evaluation of peroxyl radical scavenger capacity: Application to carotenoids and structure-activity relationship," *Food Chemistry, 135*(3), 2103–2111.
10. Nasrabadi, M. R. N., & Razavi, S. H., (2011). "Optimization of b-carotene production by a mutant of the lactosepositive yeast *Rhodotorula acheniorum* from whey ultrafiltrate," *Food Science and Biotechnology, 20*(2), 445–454.
11. Meléndez-Martínez, A. J., Vicario, I. M., & Heredia, F. J., (2004). "Nutritional importance of carotenoid pigments," *Latin American Nutrition Archives, 54*(2), 149–154.
12. Pour, H. S. R., Tavakoli, O., & Sarrafzadeh, M. H., (2016). "Experimental optimization of SC-CO2 extraction of carotenoids from *Dunaliella salina*," *The Journal of Supercritical Fluids, 121,* 89–95.
13. Hernández-Almanza, A. J., Montañez, G., Martínez, A., Aguilar-Jiménez, J., Contreras-Esquivel, C., & Aguilar, C. N., (2016). "Lycopene: Progress in microbial production," *Trends in Food Science and Technology, 56,* 142–148.
14. Valduga, E., Rausch, R. A., H. Cence, K., Colet, R., Tiggemann, L., Zeni, J., & Toniazzo, G., (2014). "Carotenoids production from a newly isolated *Sporidiobolus pararoseus* strain using agroindustrial substrates," *Biocatalysis and Agricultural Biotechnology, 3*(2), 207–213.
15. Yimyoo, T., Yongmanitchai, W., & Limtong, S., (2011). "Carotenoid production by rhodosporidium paludigenum dmku3-lpk4 using glycerol as the carbon source," *Kasetsart Journal – Natural Science, 45*(1), 90–100.
16. Silva, C. M., Borba, T. M. V., Burkert, C. A., & Burkert, J. F. M., (2012). "Carotenoid production by phaffia rhodozyma using raw glycerol as an additional carbon source," *International Journal of Food Engineering, 8*(4). 1–15, 18.
17. Macernis, M., Sulskus, J., Malickaja, S., Robert, B., & Valkunas, L., (2014). Resonance Raman spectra and electronic transitions in carotenoids. *J. Phys. Chem, A 118* (10), 1817–1825.
18. Meléndez-Martínez, A. J., Vicario, I. M., & Heredia, F. J., (2004). "Stability of carotenoid pigments in foods," *Latin American Nutrition Archives, 54*(2), 209–215.
19. Aburai, N., Sumida, D., & Abe, K., (2015). "Effect of light level and salinity on the composition and accumulation of free and ester-type carotenoids in the aerial microalga *Scenedesmus* sp. (Chlorophyceae)," *Algal Research, 8,* 30–36.

20. Jørgensen, K., & Skibsted, L., (1990). "Light sensitivity of carotenoids used as food colors," *Zeitschrift für Lebensmittel-Untersuchung und Forschung, 190,* 306–313.
21. Rao, A. V., & Rao, L. G., (2007). "Carotenoids and human health," *Pharmacological Research, 55*(3), 207–216.
22. Dong Xiao, Y., Yang, H. W., Jing, Li, D., Feng, S. J., Quan, L. C., Yu Wei, Q., et al., (2018). "Thermal degradation kinetics of all-trans and cis-carotenoids in a light-induced model system," *Food Chemistry, 239,* 360–368.
23. Hiranvarachat, B., Devahastin, S., Chiewchan, N., & Vijaya Raghavan, G. S., (2013). "Structural modification by different pretreatment methods to enhance microwave-assisted extraction of b-carotene from carrots," *Journal of Food Engineering, 115*(2), 190–197.
24. Lee, P. C., Mijts, B. N., & Schmidt-Dannert, C., (2004). "Investigation of factors influencing production of the monocyclic carotenoid torulene in metabolically engineered Escherichia coli," *Applied Microbiology and Biotechnology, 65*(5), 538–546.
25. Yin, C., Yang, S., Liu, X., & Yan, H., (2013). "Efficient extraction of astaxanthin from *Phaffia rhodozyma* with polar and non-polar solvents after acid washing," *Chinese Journal of Chemical Engineering, 21*(7), 776–780.
26. Yamaguchi, M., Hasegawa, I., Yahagi, N., Ishigaki, Y., Takano, F., & Ohta, T., (2010). "Carotenoids modulate cytokine production in Peyer's patch cells ex vivo," *Journal of Agricultural and Food Chemistry, 58*(15), 8566–8572.
27. Tan, C., Xue, J., Abbas, S., Feng, B., Zhang, X., & Xia, S., (2014). "Liposome as a delivery system for carotenoids: Comparative antioxidant activity of carotenoids as measured by ferric reducing antioxidant power, DPPH assay and lipid peroxidation," *Journal of Agricultural and Food Chemistry, 62*(28), 6726–6735.
28. Hernández-Almanza, A., Muñiz-Márquez, D. B., De La Rosa, O., Navarro, V., Martínez-Medina, G., Rodríguez-Herrera, R., et al., (2017). "Chapter 4 – Microbial production of bioactive pigments, oligosaccharides, and peptides A2 – grumezescu, alexandru mihai," In: Holban, A. M. B. T. F. B., (ed.), *Handbook of Food Bioengineering* (pp. 95–134). Academic Press.
29. Kalinowski, C. T., Socorro, J., & Robaina, L. E., (2013). "Effect of dietary canthaxanthin on the growth and lipid composition of red porgy (Pagrus pagrus)," *Aquaculture Research,* 893–900.
30. Kistler, A., Liechti, H., Pichard, L., Wolz, E., Oesterhelt, G., Hayes, A., & Maurel, P., (2002). "Metabolism and CYP-inducer properties of astaxanthin in man and primary human hepatocytes." *Archives of Toxicology, 75*(11–12), 665–675.
31. Kindlund, P. J., (2011). "Astaxanthin boosts muscle performance," *Nutrafoods, 10*(2), 49–53.
32. Capelli, B., Bagchi, D., & Cysewski, G. R., (2014). "Synthetic astaxanthin is significantly inferior to algal-based astaxanthin as an antioxidant and may not be suitable as a human nutraceutical supplement," *Nutrafoods, 12*(4), 145–152.
33. Phan-Thi, H., Durand, P., Prost, M., Prost, E., & Waché, Y., (2016). "Effect of heat-processing on the antioxidant and prooxidant activities of b-carotene from natural and synthetic origins on red blood cells," *Food Chemistry, 190,* 1137–1144.

34. Lim, J. Y., Kim, Y. S., Kim, K. M., Min, S. J., & Kim, Y., (2014). "β-carotene inhibits neuroblastoma tumorigenesis by regulating cell differentiation and cancer cell stemness," *Biochemical and Biophysical Research Communications, 450*(4), 1475–1480.
35. Verrijssen, T. A. J., Verkempinck, S. H. E., Christiaens, S., Van Loey, A. M., & Hendrickx, M. E., (2015). "The effect of pectin on in vitro β-carotene bioaccessibility and lipid digestion in low fat emulsions," *Food Hydrocolloids, 49*, 73–81.
36. Roberts, R. L., Green, J., & Lewis, B., (2009). "Lutein and zeaxanthin in eye and skin health," *Clinics in Dermatology, 27*(2), 195–201.
37. Dawczynski, J., Jentsch, S., Schweitzer, D., Hammer, M., Lang, G. E., & Strobel, J., (2013). "Long term effects of lutein, zeaxanthin and omega-3-LCPUFAs supplementation on optical density of macular pigment in AMD patients: The LUTEGA study," *Graefe's Archive for Clinical and Experimental Ophthalmology, 251*(12), 2711–2723.
38. Song, J. F., Li, D. J., Pang, H. L., & Liu, C. Q., (2015). "Effect of ultrasonic waves on the stability of all-trans lutein and its degradation kinetics," *Ultrasonics Sonochemistry. 27, 602–608.
39. Zhou, X., Zhang, F., Hu, X., Chen, J., Wen, X., Sun, Y., et al., (2015). "Inhibition of inflammation by astaxanthin alleviates cognition deficits in diabetic mice," *Physiology and Behavior, 151*, 412–420.
40. Lakshminarayana, R., Sathish, U. V., Dharmesh, S. M., & Baskaran, V., (2010). "Antioxidant and cytotoxic effect of oxidized lutein in human cervical carcinoma cells (HeLa)," *Food and Chemical Toxicology, 48*(7), 1811–1816.
41. Ozawa, Y., & Sasaki, M., (2014). "Lutein and oxidative stress-mediated retinal neurodegeneration in diabetes," In: Elsevier (ed.), *Diabetes: Oxidative Stress and Dietary Antioxidants* (pp. 223–229).
42. Berrow, E. J., Bartlett, H. E., & Eperjesi, F., (2011). "Do lutein, zeaxanthin and macular pigment optical density differ with age or age-related maculopathy?," *e-SPEN, 6*(4), 197–201.
43. Maci, S., & Santos, R., (2015). "The beneficial role of lutein and zeaxanthin in cataracts," *Nutrafoods, 14*(2), 63–69.
44. Kijlstra, A., Tian, Y., Kelly, E., R., & Berendschot, T. T. J. M., (2012). "Lutein: More than just a filter for blue light," *Progress in Retinal and Eye Research, 31*(4), 303–315.
45. Ravi, K. B., Raghunatha, R. K. R., Shankaranarayanan, J., Deshpande, J. V., Juturu, V., & Soni, M. G., (2014). "Safety evaluation of zeaxanthin concentrate (OmniXan™): Acute, subchronic toxicity and mutagenicity studies," *Food and Chemical Toxicology, 72*, 30–39.
46. Verrijssen, T. A. J., Smeets, K. H. G., Christiaens, S., Palmers, S., Van Loey, A. M., & Hendrickx, M. E., (2015). "Relation between in vitro lipid digestion and β-carotene bioaccessibility in β-carotene-enriched emulsions with different concentrations of L-α-phosphatidylcholine," *Frin, 67*, 60–66.
47. Agarwal, S., & Rao, A. V., (2000). "Tomato lycopene and its role in human health and chronic diseases," *Cmaj, 163*(6), 739–744.
48. Yang, P. M., Wu, Y. Z. Z., Zhang, Q., & Wung, B. S., (2016). "Lycopene inhibits ICAM-1 expression and NF-κB activation by Nrf2-regulated cell redox state in human retinal pigment epithelial cells," *Life Sciences, 155*, 94–101.

49. Campos, K. K. D., Araújo, G. R., Martins, T. L., Bandeira, A. C. B., De, G., Costa, P., et al., (2017). "The antioxidant and anti-inflammatory properties of lycopene in mice lungs exposed to cigarette smoke," *The Journal of Nutritional Biochemistry, 48*, 9–20.
50. Sahin, K., Cross, B., Sahin, N., Ciccone, K., Suleiman, S., Osunkoya, A. O., et al., (2015). "Lycopene in the prevention of renal cell cancer in the TSC2 mutant Eker rat model," *Archives of Biochemistry and Biophysics, 572*, 36–39.
51. Boyacioglu, M., Kum, C., Sekkin, S., Yalinkilinc, H. S., Avci, H., Epikmen, E., T., et al., (2016). "The effects of lycopene on DNA damage and oxidative stress on indomethacin-induced gastric ulcer in rats," *Clinical Nutrition, 35*(2), 428–435.
52. Feofilova, E. P., Tereshina, V. M., Memorskaia, A. S., Dul'kin, L. M., & Goncharov, N. G., (2006). "Fungal lycopene: The biotechnology of its production and prospects for its application in medicine," *Mikrobiologiia, 75*(6), 725–730.
53. Erdman, J. W., Ford, N. A., & Lindshield, B. L., (2009). "Are the health attributes of lycopene related to its antioxidant function?" *Archives of Biochemistry and Biophysics, 483*(2), 229–235.
54. Vidolova, A., Bulgarian, P., & Popova, A. V., (2017). "Spectral characteristics and solubility of beta-carotene and zeaxanthin in different solvents." *Proceedings of the Bulgarian Science Website l'Acad, 70(1),* 53–60.
55. Zhou, C. H., Xu, C. J., De Sun, C., Li, X., & Chen, K. S., (2007). "Carotenoids in white- and red-fleshed loquat fruits," *Journal of Agricultural and Food Chemistry, 55*(19), 7822–7830.
56. Jáuregui, M. E. C., De La Concepción, C. C. M., & Romo, F. P. G., (2011). "Carotenoids and their antioxidant function: Review," *Latin American Nutrition Archives, 61*(3), 233–241.
57. Woodside, J. V., McGrath, A. J., Lyner, N., & McKinley, M. C., (2015). "Carotenoids and health in older people," *Maturitas, 80*(1), 63–68.
58. Stigliani, A. L., Giorio, G., & D'Ambrosio, C., (2011). "Characterization of P450 carotenoid B- and E-hydroxylases of tomato and transcriptional regulation of xanthophyll biosynthesis in root, leaf, petal and fruit," *Plant and Cell Physiology, 52*(5), 851–865.
59. Frede, K., Henze, A., Khalil, M., Baldermann, S., Schweigert, F. J., & Rawel, H., (2014). "Stability and cellular uptake of lutein-loaded emulsions," *Journal of Functional Foods, 8*(1), 118–127.
60. Maiani, G., Castón, M. J. P., Catasta, G., Toti, E., Cambrodón, I. G., Bysted, A., et al., (2009). "Carotenoids: Actual knowledge on food sources, intakes, stability and bioavailability and their protective role in humans," *Molecular Nutrition and Food Research, 53*(2), 194–218.
61. Cazzonelli, C. I., (2011). "Carotenoids in nature: Insights from plants and beyond," *Functional Plant Biology, 38*(11), 833–847.
62. Polívka, T., & Frank, H. A., (2010). "Molecular factors controlling photosynthetic light harvesting by carotenoids," *Accounts of Chemical Research, 43*(8), 1125–1134.
63. Skibsted, L. H., (2012). "Carotenoids in antioxidant networks. Colorants or radical scavengers," *Journal of Agricultural and Food Chemistry, 60*(10), 2409–2417.
64. Britton, G., (2008). "Functions of intact carotenoids," in *Carotenoids, 4*, 189–212.

65. Frank, H. A., & Brudvig, G. W., (2004). "Current topics carotenoids in photosynthesis," *Biochemistry*, *43*(27), 8607–8615.
66. Isaksson, C., (2009). "The chemical pathway of carotenoids: From plants to birds," *Ardea*, *97*(1), 125–128.
67. Fraser, P. D., Enfissi, E. M. A., & Bramley, P. M., (2009). "Genetic engineering of carotenoid formation in tomato fruit and the potential application of systems and synthetic biology approaches," *Archives of Biochemistry and Biophysics*, *483*(2), 196–204.
68. Koul, A., Yogindran, S., Sharma, D., & Kaul, S., (2016). "Plant physiology and biochemistry carotenoid pro fi ling, in silico analysis and transcript pro fi ling of miRNAs targeting carotenoid biosynthetic pathway genes in different developmental tissues of tomato," *Plant Physiology et Biochemistry*, *108*, 412–421.
69. Valduga, E., Valério, A., Treichel, H., Júnior, A. F., & Di, M. L., (2009). "Optimization of the production of total carotenoids by Sporidiobolus salmonicolor (CBS 2636) using response surface technique," *Food and Bioprocess Technology*, *2*(4), 415–421.
70. Buzzini, P., Martini, A., Gaetani, M., Turchetti, B., Pagnoni, U. M., & Davoli, P., (2005). "Optimization of carotenoid production by Rhodotorula graminis DBVPG 7021 as a function of trace element concentration by means of response surface analysis," *Enzyme and Microbial Technology*, *36*(5–6), 687–692.
71. Park, P. K., Cho, D. H., Kim, E. Y., & Chu, K. H., (2005). "Optimization of carotenoid production by *Rhodotorula glutinis* using statistical experimental design," *World Journal of Microbiology and Biotechnology*, *21*(4), 429–434.
72. Rodrigues, D. B., Flores, É. M. M., Barin, J. S., Mercadante, A. Z., Jacob-Lopes, E., & Zepka, L. Q., (2014). "Production of carotenoids from microalgae cultivated using agroindustrial wastes," *Food Research International*, *65*, 144–148.
73. Hernández-Almanza, A., Montañez-Sáenz, J., Martínez-Ávila, C., Rodríguez-Herrera, R., & Aguilar, C. N., (2014). "Carotenoid production by *Rhodotorula glutinis* YB-252 in solid-state fermentation," *Food Bioscience*, *7*, 31–36.
74. López-Nieto, M. J., Costa, Peiro, J. E., Méndez, E., Rodríguez-Sáiz, M., De La Fuente, J. L., Cabri, W., et al., (2004). "Biotechnological lycopene production by mated fermentation of Blakeslea trispora," *Applied Microbiology and Biotechnology*, *66*(2), 153–159.
75. Choudhari, S. M., Ananthanarayan, L., & Singhal, R. S., (2008). "Use of metabolic stimulators and inhibitors for enhanced production of β-carotene and lycopene by *Blakeslea trispora* NRRL 2895 and 2896," *Bioresource Technology*, *99*(8), 3166–3173.
76. Frengova, G. I., & Beshkova, D. M., (2009). "Carotenoids from Rhodotorula and Phaffia: Yeasts of biotechnological importance," *Journal of Industrial Microbiology and Biotechnology*, *36*(2), 163–180.
77. Aksu, Z., & Eren, A. T., (2007). "Production of carotenoids by the isolated yeast of Rhodotorula glutinis," *Biochemical Engineering Journal*, *35*(2), 107–113.
78. Cardoso, L. A. C., Jäckel, S., Karp, S. G., Framboisier, X., Chevalot, I., & Marc, I., (2016). "Bioresource technology improvement of sporobolomyces ruberrimus carotenoids production by the use of raw glycerol," *Bioresource Technology*, *200*, 374–379.

79. Moroni, S. C., De Matos de Borba, T., Juliano, K. S., & De Medeiros, B. J. F., (2016). "Raw glycerol and parboiled rice effluent for carotenoid production: Effect of the composition of culture medium and initial pH," *Food Technology and Biotechnology*, *54*(4), 489–496.
80. Khanafari, A., (2008). "Light requirement for the carotenoids production by Mucor hiemalis," *Iranian Journal of Basic Medical Sciences*, *11*, 25–32.
81. Ziegelhoffer, E. C., & Donohue, T. J., (2009). "Bacterial responses to photo-oxidative stress," *Nature Reviews Microbiology*, *7*(12), 856–863.
82. Zhang, Z., Zhang, X., & Tan, T., (2014). "Lipid and carotenoid production by Rhodotorula glutinis under irradiation/high-temperature and dark/Low-temperature cultivation," *Bioresource Technology*, *157*, 149–153.
83. Cheng, Y. T., & Yang, C. F., (2016). "Using strain Rhodotorula mucilaginosa to produce carotenoids using food wastes," *Journal of the Taiwan Institute of Chemical Engineers*, *61*, 270–275.
84. Bhosale, P. B., & Gadre, R. V., (2001). "Production of β-carotene by a mutant of *Rhodotorula glutinis*," *Applied Microbiology and Biotechnology*, *55*(4), 423–427.
85. Adil, I. H., Çetin, H. I., Yener, M. E., & Bayindirli, A., (2007). "Subcritical (carbon dioxide + ethanol) extraction of polyphenols from apple and peach pomaces, and determination of the antioxidant activities of the extracts," *Journal of Supercritical Fluids*, *43*(1), 55–63.
86. Davarnejad, R., Kassim, K. M., Zainal, A., & Sata, S. A., (2008). "Supercritical fluid extraction of b-carotene from crude palm oil using CO_2," *Journal of Food Engineering*, *89*(4), 472–478.
87. Michelon, M., De Matos de Borba, T., Da Silva Rafael, R., Burkert, C. A. V., & De Medeiros, B. J. F., (2012). "Extraction of carotenoids from *Phaffia rhodozyma*: A comparison between different techniques of cell disruption," *Food Science and Biotechnology*, *21*(1), 1–8.
88. Yolmeh, M., Habibi, N. M. B., & Farhoosh, R., (2014). "Optimization of ultrasound-assisted extraction of natural pigment from annatto seeds by response surface methodology (RSM)," *Food Chemistry*, *155*, 319–324.
89. Cardenas-Toro, F. P., Alcázar-Alay, S. C., Coutinho, J. P., Godoy, H. T., Forster-Carneiro, T., & Meireles, M. A. A., (2015). "Pressurized liquid extraction and low-pressure solvent extraction of carotenoids from pressed palm fiber: Experimental and economical evaluation," *Food and Bioproducts Processing*, *94*, 90–100.
90. Prado, I. M., Prado, G. H. C., Prado, J. M., & Meireles, M. A. A., (2013). "Supercritical CO_2 and low-pressure solvent extraction of mango (Mangifera indica) leaves: Global yield, extraction kinetics, chemical composition and cost of manufacturing," *Food and Bioproducts Processing*, *91*(4), 656–664.
91. Fornasari, C. H., Secco, D., Santos, R. F., Da Silva, T. R. B., Galant, L. N. B., Tokura, L. K., et al., (2017). "Efficiency of the use of solvents in vegetable oil extraction at oleaginous crops," *Renewable and Sustainable Energy Reviews*, *80*, 121–124.
92. Kwang, H. C., Lee, H. J., Koo, S. Y., Song, D. G., Lee, D. U., & Pan, C. H., (2010). "Optimization of pressurized liquid extraction of carotenoids and chlorophylls from chlorella vulgaris," *Journal of Agricultural and Food Chemistry*, *58*(2), 793–797.

93. Mustafa, A., Trevino, L. M., & Turner, C., (2012). "Pressurized hot ethanol extraction of carotenoids from carrot by-products," *Molecules, 17*(2), 1809–1818.
94. Mustafa, A., & Turner, C., (2011). "Pressurized liquid extraction as a green approach in food and herbal plants extraction: A review," *Analytica Chimica Acta, 703*(1), 8–18.
95. Amosova, A. S., Ivakhnov, A. D., Skrebets, T. E., Ulyanovskiy, N. V., & Bogolitsyn, K. G., (2014). "Supercritical fluid extraction of carotenoids from shantane carrot," *Russian Journal of Physical Chemistry B, 8*(7), 963–966.
96. Catchpole, O. J., Tallon, S. J., Eltringham, W. E., Grey, J. B., Fenton, K. A., Vagi, E. M., et al., (2009). "The extraction and fractionation of specialty lipids using near critical fluids," *Journal of Supercritical Fluids, 47*(3), 591–597.
97. Macías-Sánchez, M. D., Mantell, C., Rodríguez, M., Martínez de la Ossa, E., Lubián, L. M., & Montero, O., (2009). "Comparison of supercritical fluid and ultrasound-assisted extraction of carotenoids and chlorophyll a from Dunaliella salina," *Talanta, 77*(3), 948–952.
98. Hiranvarachat, B., & Devahastin, S., (2014). "Enhancement of microwave-assisted extraction via intermittent radiation: Extraction of carotenoids from carrot peels," *Journal of Food Engineering, 126,* 17–26.
99. Ho, K. K. H. Y., Ferruzzi, M. G., Liceaga, A. M., & San Martín-González, M. F., (2015). "Microwave-assisted extraction of lycopene in tomato peels: Effect of extraction conditions on all-trans and cis-isomer yields," *LWT – Food Science and Technology, 62*(1), 160–168.
100. De Medeiros, F. O., Alves, F. G., Lisboa, C. R., Martins, D. D. S., Burkert, C. A., V., & Kalil, S. J., (2008). "Ondas ultrassónicas e pérolas de vidro: Um novo método de extracao de b-galactosidase para uso em laboratório," *Quimica Nova, 31*(2), 336–339.
101. Ordóñez-Santos, L. E., Pinzón-Zarate, L. X., & González-Salcedo, L. O., (2015). "Optimization of ultrasonic-assisted extraction of total carotenoids from peach palm fruit (Bactris gasipaes) by-products with sunflower oil using response surface methodology," *Ultrasonics Sonochemistry, 27,* 560–566.
102. Chemat, F., Zill-E-Huma, & Khan, M. K., (2011). "Applications of ultrasound in food technology: Processing, preservation and extraction," *Ultrasonics Sonochemistry, 18*(4), 813–835.
103. Dey, S., & Rathod, V. K., (2013). "Ultrasound assisted extraction of β-carotene from Spirulina platensis," *Ultrasonics Sonochemistry, 20*(1), 271–276.
104. Eh, A. L. S., & Teoh, S. G., (2012). "Novel modified ultrasonication technique for the extraction of lycopene from tomatoes," *Ultrasonics Sonochemistry, 19*(1), 151–159.
105. Luterotti, S., Marković, K., Franko, M., Bicanic, D. Madžgalj, A., & Kljak, K., (2013). "Comparison of spectrophotometric and HPLC methods for determination of carotenoids in foods." *Food Chemistry, 140*(1–2), 390–397.
106. Valdivielso, I., Bustamante, M. Á., Ruiz de Gordoa, J. C., Nájera, A. I., De Renobales, M., & Barron, L. J. R., (2015). "Simultaneous analysis of carotenoids and tocopherols in botanical species using one step solid–liquid extraction followed by high-performance liquid chromatography," *Food Chemistry, 173,* 709–717.

107. Medina-Molotla, N., Thorsteinsdóttir, H., Frixione, E., & Kuri-Harcuch, W., (2017). "Some factors limiting transfer of biotechnology research for health care at Cinvestav: A Mexican scientific center," *Technology in Society*, *48*, 1–10.
108. Espinoza-hernández, T. C., Rodríguez-herrera, R., Aguilar, C. N., Lara-victoriano, F., Reyes-valdés, M. H., & Reyes, C., (2013). "Characterization of three novel pigment-producing Penicillium strains isolated from the Mexican semi- desert," *African Journal of Biotechnology*, *12*(22), 3405–3413.
109. Aguilar-Machado, D., Morales-Oyervides, L., Contreras-Esquivel, J. C., Aguilar, C., Méndez-Zavala, A., Raso, J., et al., (2017). "Effect of ohmic heating processing conditions on color stability of fungal pigments," *Food Science and Technology International*, 1–11.
110. Morales-Oyervides, L., Oliveira, J. C., Sousa-Gallagher, M. J., Méndez-Zavala, A., & Montañez, J. C., (2015). "Effect of heat exposure on the color intensity of red pigments produced by Penicillium purpurogenum GH2," *Journal of Food Engineering*, *164*, 21–29.
111. Mata-Gómez, L. C., (2015). "Production and characterization of carotenoids by Rhodotorula sp. in submerged culture using an agro-industrial residue as culture medium" Autonomous University of Coahuila.

CHAPTER 4

HYDROGELS OF BIOPOLYMERS FUNCTIONALIZED WITH BIOACTIVE SUBSTANCES AS A COATING FOR FOOD PRESERVATION

ALEJANDRA ISABEL VARGAS-SEGURA,
MÓNICA LIZETH CHÁVEZ-GONZÁLEZ,
JOSÉ LUIS MARTÍNEZ-HERNÁNDEZ,
RODOLFO RAMOS-GONZÁLEZ, ANNA ILINÁ,
and ELDA PATRICIA SEGURA-CENICEROS

Autonomous University of Coahuila. School of Chemistry, Blvd. Venustiano Carranza, Col. República Oriente, C.P. 25000 Saltillo, Coahuila, Mexico, E-mail: psegura@uadec.edu.mx

ABSTRACT

In recent years, due to new habits of life, there is a growing concern of consumers for their health and the effect of diet as a risk factor for diseases such as obesity, coronary heart disease, hypercholesterolemia, etc. Due to the above, the production and consumption of minimally processed fruits and vegetables are becoming more popular in the market, since these are natural products with numerous health benefits and allow time savings associated with washing, peeling and cut from these foods. Thus, these minimally processed products require a high sanitary control, being the storage the most critical period, since it can suffer attacks of microorganisms or deterioration of the product. The use of active packaging can retain undesirable substances from the product or the environment (O_2, H_2O, CO_2, ethylene, etc.) or release beneficial substances into the product such as antioxidants, antimicrobials, etc., which can be carried out from the material of packaging. Hydrogels are an option because of their adequate physical properties. Hydrogels are hydrophilic polymeric structures capable of absorbing a large amount of water or biological fluids. Hydrogels are of great interest as they can be designed for specific applications such as food additives, transport, and release of bioactive substances, films for the protection and coating of food (packaging material), superabsorbent water gels, chemical traps, drug vehicles, synthetic fabric construction, enzyme immobilizing agents.

4.1 INTRODUCTION

4.1.1 CONSERVATION OF FOODS

Currently, there is a concern for health and the effect of diet as a risk factor for diseases such as obesity, coronary heart disease, hypercholesterolemia, etc. Due to this, the production and consumption of minimally processed foods are more popular in the market, since these are natural products with numerous health benefits and also allow time savings associated with washing, peeling and cutting. However, in order to achieve this, a high sanitary control is required after harvest, with packing being the most critical period, as there may be attacks of microorganisms and reactions of deterioration of the food as a result of respiration and maturation of the product [1, 2].

In order to increase the shelf life of foods and to reduce the losses caused by microorganism's growth, there are different packaging techniques, including modified atmosphere packaging (MAP), coating the fruit with edible films [1] and active packaging [2]. Minimally processed vegetable products are those that have undergone a series of mild treatments and maintain a fresh, uniform and consistent product quality, compared to the initial product. Fresh cut fruits (FFC) are products prepared from basic operations such as washing, peeling, deboning and cutting, and may include sanitizing operations with chlorinated derivatives, hydrogen peroxide, ozone, natural antimicrobials and others, treatments with stabilizers such as ascorbic acid, or firmness retainers, such as calcium salts [3]. These products should be stored at refrigeration temperatures (2–5°C) to prevent deterioration, as well as the growth of microorganisms, and if possible, packaging with a protective, modified atmosphere or active packaging. In this way, FFC can be consumed for a period between 7 and 14 days depending on the product and conservation technique applied [4].

The packaging of food has a fundamental role during its commercialization, as it is the main bringer between the environment and the product. Foods deteriorate over time, through the action of living organisms (molds, bacteria, insects, rodents, etc.), and/or the physicochemical action of the environment (temperature, relative humidity, oxygen, radiation, etc.) of the food itself. Therefore, the packaging used for its conservation should reduce the effect of external factors, protecting the integrity of the product and avoiding or delaying the deterioration of nutritional, sensory and sanitary characteristics that define its quality and acceptance for consumption. Containers of plastic materials are generally not biodegradable and the raw materials used in their manufacture and/or processing may be carcinogenic. Thus, natural resources as a source of conservation and recycling are an excellent option for innovation in the development of new biodegradable products, since its total biodegradation in CO_2, water, and organic fertilizer is a great advantage over synthetic resources [5]. Natural biopolymers used in packaging or food coating come from four major sources: Animal origin (collagen/gelatin), marine origin (chitin/chitosan), agricultural origin (lipids and fats and hydrocolloids: Proteins and polysaccharides) and microbial origin (polylactic acid (PLA), polyhydroxyalkanoates (PHA)) [6]. The so-called biodegradable edible coatings of natural biopolymers must slow the loss of water and weight of food, avoid or reduce gaseous exchange, maintain quality, with no pH changes, maintain skin firmness and minimize respiration rate in fruits and vegetables, which leads to fermentation and rot [7–9]. The coatings used may be of different types, such as polysaccharides, proteins and lipids. They should be edible, with good aesthetic appearance, a good barrier to oxygen and water vapor [10].

Active packaging is a packaging technology whose objective is to maintain or extend the shelf life and quality of the product by two techniques, namely the release of active substances or the removal of undesirable components [11, 12]. These active containers can retain undesirable substances of the product or the environment (O_2, H_2O, CO_2, ethylene, etc.) or release beneficial substances to the product or environment such as antioxidants, antimicrobials, etc. The release of the active substances, or retention, can be carried out from the packaging material itself, regardless of its origin, or from various devices or releasing devices which would be included within the primary packaging where the product is placed. In this context, so-called hydrogels have great applicability due to their adequate physical properties. Due to the predominantly volatile and hydrophobic nature of the antimicrobials most commonly used in active packaging [13, 14], their incorporation into these hydrogels is necessary.

4.1.2 HYDROGELS

Hydrogels are hydrophilic three-dimensional polymer networks that are able to take up large amounts of water or physiological fluid maintaining their internal network structure. These are obtained by simultaneous polymerization and cross-linking of one or more polyfunctional monomers. Hydrogels are very active from the osmotic point of view, so that, physically, they are intermediate species between the solid state and the liquid state. Interpenetrated network hydrogels (IPN) are currently being studied for their potential application in several areas [15]. The physical properties of hydrogels depend mainly on the balance of ionic groups present, generating attraction or repulsion interactions between their chains. The ionic strength of the medium where they are found, and the elasticity of the polymer network, are also important factors for their behavior [16].

Hydrogels are a network of polymers capable of absorbing large amounts of water and other solutions or mixtures. In this network there are hydrophilic groups that hydrate in an aqueous environment, giving rise to the own hydrogel structure [17]. For the formation of the hydrogel to take place, a cross-linking phenomenon must occur, which allows to avoid the dissolution of the polymer chains in the aqueous phase after the formation of the latter. Cross-linking consists of forming links between the various chains of the polymer network, either chemical (covalent), or physical (hydrogen bridges and van der Waals forces) [18]. This can be achieved by using different methods, such as varying the temperature and stirring time of the mixtures or by freeze-thaw cycles, to achieve physical cross-linking, or by adding certain additives to the mixtures, as organic acids (citric acid, maleic acid, etc.) to produce chemical cross-linking [19, 20] through the formation of ester-like bonds. Because of their porous matrix and their high capacity for absorption of water or aqueous solutions, hydrogels are of high interest as controlled release devices for bioactive substances, and may even respond to external stimuli, such as changes in temperature [21, 22] or in pH.

The integrity of the polymer network is due to the presence of chemical and/or physical crosslinks between the macromolecular chains. Depending on the case, they are also considered as intelligent polymer systems or polymers sensitive to different stimuli, that is, as polymeric structures that in response to slight changes in their environment (temperature, pH, light, electric or magnetic field, ionic concentration, biological molecules, etc.) undergo changes in its structure and properties from which its applications are derived [22]. An interesting feature of many hydrogels sensitive to these stimuli is that the mechanism that causes structural change is completely reversible. These materials are considered to be of great interest as they can be designed for specific applications such as food additives, transport, and release of bioactive substances, films for the protection and coating of food (packaging material), superabsorbent water gels, chemical traps, vehicles Drug manufacturing, synthetic fabric construction, enzyme immobilizing agents, etc. If these structures are used as reservoirs and transport of pharmacological drugs, hydrogels can be used as drug release agents at specific sites. Polymeric solids are especially suitable for forming gels thanks to their long chain structure. The flexibility of these chains makes it possible for them to deform to allow the entry of solvent molecules into their three-dimensional structure. There is a direct relationship between the properties of a hydrogel and its structure. Both characteristics cannot be considered in isolation, since the method of synthesis has a decisive influence on them. Therefore, when the properties of the hydrogels are exposed, reference must be made to the structural parameters that condition them. There are various types of polymers capable of forming hydrogels, obtained by chemical synthesis or from natural sources and, depending on their nature, they may be biocompatible and/or biodegradable [23].

In the synthesis of hydrogels, the cross-linking agent is responsible for the reticulated structure of the hydrogel by having more than two reactive groups in its structure, being usual tetrafunctional compounds, and even hexafunctionals [24]. Hydrogels may be synthesized from natural or synthetic polymers; giving physical and chemical hydrogels; which can have a heterogeneous organization of independent domains forming a network.

Physical hydrogels are organized in distinct domains formed by molecular entanglements, free chain ends, held together by weak hydrophobic associations, ionic interactions, or hydrogen bonding [25, 26]. Also, called 'reversible' or 'pseudo' gels, exhibit high water sensitivity (degrade and even disintegrate completely in water) and thermo-reversibility (melt to a polymer solution when exposed to heat).

Chemical hydrogels (also called 'irreversible' or 'permanent' gels) are networks of polymer chains covalently linked (Figure 4.1). The cross-linking is carried out by chemical reactions with aldehydes [27] or radiation (e.g., electron beam exposure, gamma-radiation, or UV light) [27, 28]. Uneven distribution of cross-linking within the gel leads to the development of some zones in which typical 'reversible' features are still dominant and other zones with permanent properties arising from the cross-linked network. Chemical hydrogels neither disintegrate nor dissolve in aqueous solutions. The chemical hydrogel hydrate and swell until an equilibrium state is reached, which in turn depends on the grade of the cross-linking. The swelling process is governed first by the water binding at the hydrophilic sites of the biomolecules, followed by the entrapment in the gel network of one or more hydration layers which form a 'shell like' structure around the biomacromolecules.

The hydrogels are polymer networks extensively swollen with water. Hydrophilic gels that are usually referred to as hydrogels are networks of polymer chains that are sometimes found as colloidal gels in which water is the dispersion medium. Hydrogels can be defined in different ways:

1. Hydrogel is a water-swollen, and cross-linked polymeric network produced by the simple reaction of one or more monomers.
2. Another definition is that it is a polymeric material that exhibits the ability to swell and retain a significant fraction of water within its structure, but will not dissolve in water.

They possess also a degree of flexibility very similar to natural tissue due to their large water content. The ability of hydrogels to absorb water arises from hydrophilic functional groups attached to the polymeric backbone, while their resistance to dissolution arises from cross-links between network chains. Recently, hydrogels have been defined as two- or multi-component systems consisting of a three-dimensional network of polymer chains and water that fills the space between macromolecules. Depending on the properties of the polymer (polymers) used, as well as on the nature and density of the network joints, such structures in an equilibrium can contain various amounts of water; typically, in the swollen state the mass fraction of water in a hydrogel is much higher than the mass fraction of polymer [29].

With the establishment of the first synthetic hydrogels, the hydrogel technologies may be applied to hygienic products, agriculture, drug delivery systems, sealing, coal dewatering, artificial snow, food additives, pharmaceuticals, biomedical applications tissue engineering and regenerative medicines, diagnostics, wound dressing, separation of biomolecules or cells and barrier materials to regulate biological adhesions, and biosensor. The hydrogels used in tissue engineering have the functionality of allowing that the live cells are adequately accommodated or designed to dissolve or degraded, act as a guide for the "in vitro" and "in vivo" tissue development, have the ability to retain water in their structure in the same way they retain bioactive proteins. In particular, polysaccharides in general are non-toxic, highly biocompatible and with physicochemical characteristics that make them suitable for different biomedical and pharmaceutical applications, including drug delivery systems. In addition, they are abundant and are available from renewable sources such as algae, plants and cultures of selected microbial strains. They have, therefore, a great variety of composition and properties that cannot be easily imitated in the laboratory with synthetic polymers, being their more economic production [30].

Some of the biodegradable and biocompatible polymers that can be used to form hydrogels are given in the following subsections.

4.1.2.1 STARCH

Starch is one of the most abundant polysaccharides, only behind cellulose, which is renewable, cheap and widely available [31]. This polymer is formed, in the native state, by two macromolecular components: Amylose and amylopectin. Amylose is a linear polymer formed by glucose units linked by α-1,4 bonds. In contrast, amylopectin is a highly-branched molecule consisting of short chains of α-1,4-linked α-1,4-linked linkages with α-1,6 branching occurring every 25–30 units of glucose [31]. The main sources for commercial starch production are potato, wheat, maize and rice. Most of these starches contain 20 to 30% amylose, the balance being amylopectin and minor components (less than 1%) such as lipids and proteins. Starches from other sources such as cassava, amaranth, mango and banana have also been used commercially and/or studied, among others. Starch is an interesting biopolymer for the formation of hydrogels due to its abundance in nature, biodegradability, biocompatibility and its low cost in relation to other polymers [32].

4.1.2.2 POLI(VINYL ALCOHOL) (PCA)

PVA, although obtained after a synthesis process, is fully biodegradable and its compatibility with starch to form hydrogels has been tested previously [33]. Polyvinyl alcohol (PVA) is widely used in biotechnology to encapsulate enzymes and microorganisms and, presently, in the pharmaceutical industry as a slow release drug carrier. The chemical structure is presented in Figure 4.2.

PVA is industrially produced by polymerization of vinyl acetate to polyvinyl acetate (PVAc), followed by hydrolysis to PVA with the release of acetate groups. This conversion is not complete, and the degree of hydrolysis achieved must be determined. For this reason, PVA should always be considered as a copolymer of PVA and PVAc. The degree of hydrolysis or the content of acetate groups has strong implications on the

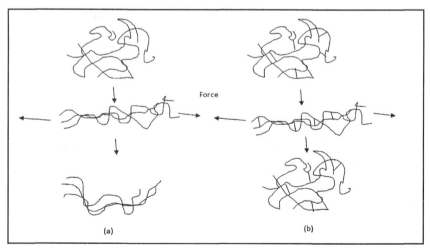

FIGURE 4.1 (a) Physical hydrogels and (b) Chemical hydrogels [28].

chemical properties, solubility and crystallization of the polymer. PVA can be cross-linked in various ways, one of which is through agents which generate covalent bonds between different polymer chains, such as glutaraldehyde, acetaldehyde, formaldehyde and other monoaldehydes in the presence of sulfuric, acetic or methanol. Gamma-ray irradiation, having as the main advantage the absence of contaminants in the structure obtained; or by physical interactions between strings. Physical hydrogels are formed from hydrogen bridge-type interactions. These generate zones of crystalline character that provide sites with structural rigidity from which the polymer matrix develops. Aqueous solutions of PVA have the characteristic of generating large numbers of crystalline nuclei, which provides the matrix with certain properties, such as high degree of swelling, elastic nature and high mechanical strength. Systems are not yet well characterized in the formation of mixtures [29].

Similarly, hydrogels can be formed from starch and PVA, with or without addition of polycarboxylic acids, which have a high-water retention capacity [34]. This property can be used to incorporate into the hydrogel substances capable of acting as antimicrobials, thus limiting the growth of bacteria and fungi responsible for the deterioration of food. One of the most commonly used antimicrobial substances in new packaging applications is essential oils. These substances, extracted from different plant species generally contain a large number of phenolic compounds and terpenes which confer a high antimicrobial and antioxidant capacity [35].

4.1.2.3 GELATIN

Because of its programmable, biodegradable and reabsorbable nature, proteins are excellent candidates for the development of biomaterials [10]. A classic protein hydrogel builder is gelatin, which comes from animal collagen by treatment at extreme pH followed by a warming up [36, 37]. Gelatin is a generic term for a mixture of purified protein fractions obtained by partial acid hydrolysis (gelatin type A) or partial alkaline hydrolysis (gelatin type B) from animal collagen. Below 25°C, an aqueous solution of gelatin solidifies due to the formation of triple helices and a rigid three-dimensional network. When the temperature rises above about 30°C, changes in propeller confirmation to a more flexible structure cause the gel to become liquid again. As opposed thermal behavior is required for biomedical applications, researchers have combined gelatin with other polymers that show thermal gelling closer to body temperature (Figure 4.3).

4.1.2.4 CHITOSAN

The cross-linked chitosan hydrogels are classified as ionic and covalent hydrogels. The latter are divided into three groups: Chitosan cross-linked with itself, hybrid polymer networks and interpenetrated polymer networks. Cross-linking naturally involves two structural units, which may or may not belong to the same polymer chain. The main interactions that occur in this type of hydrogel correspond to covalent bonds; but other interactions may also occur such as hydrogen bonds and hydrophobic

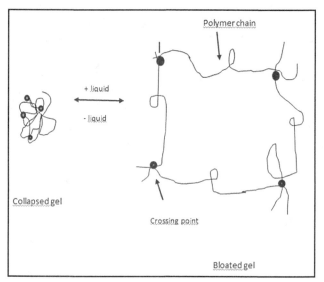

FIGURE 4.2 Chemical structure of PVA [33].

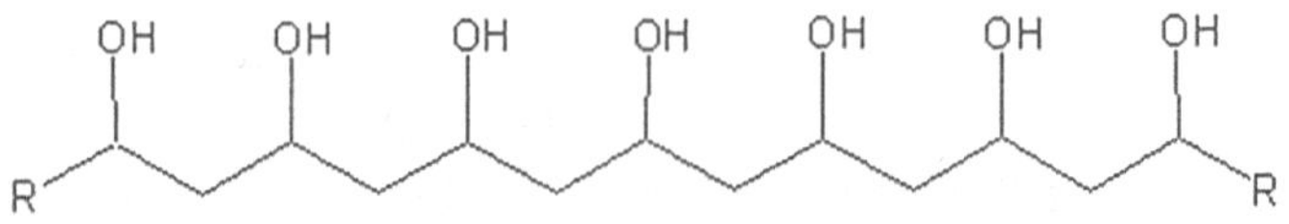

FIGURE 4.3 Scheme of hydrogel in the swollen state [29].

interactions, which are formed between chitosan units. Because chitosan has a cationic character, it can maintain a positive charge at acidic pH, giving it the ability to interact with cellular surfaces and plasma protein [21, 38].

Chitosan is an organic compound, derived from chitin, a protein present in aquatic organisms; Is a natural polymer and possesses excellent properties as antifungal, antiviral, biocompatible, biodegradable, antimicrobial, non-toxic, emulsifier, grease absorbent and contaminating metal absorber; All these properties make it considered of great application in different fields of research, in the pharmaceutical industry, cosmetics, medicine, biotechnology, food and agriculture. Chitosan presents a potential and interesting economic value, due to its versatile biological activities and chemical applications; is biodegradable and biocompatible, since it does not produce a response of the immune system and is not toxic, characteristics that, together with its polycatalytic nature, have stimulated its use in several biomedical applications [17, 33, 39]. Chitosan can form hydrogels with antibacterial and antifungal properties; therefore, hydrogel coatings used as part of the packaging may inhibit the growth of bacteria and fungi in seasonal fruits such as apple, pear and grape [33].

4.1.2.5 PECTIN

Low-methoxyl (LM) pectins provide the charge interactions required for hydrogels. LM pectins are linear polyanions (polycarboxylates) of about 300–1000 saccharide units, with polymer molecular weights ranging from 50,000 to 150,000 Da. Dissolved LM pectins do not form gel because ionized carboxylate groups (eCOO) along the molecule repel each other. In addition, each pectin chain is strongly hydrated, particularly around the carboxylate groups [22]. However, in the presence of divalent cations (usually calcium), ionic interactions between the divalent cation and carboxylate groups in the pectin chains promote the formation of junction zones between molecules in close proximity. Cross-linking and gel strength increase directly with calcium concentration and inversely with DE.

Hydrogels can offer new opportunities for the design of efficient packaging materials with desirable properties (i.e., durability, biodegradability mechanical properties etc.) [23, 24]. Packaging material for food industry is very authoritative because it needs to fulfill several objectives: Physical protection, barrier protection, information transmission, marketing communication and so on [25]. Beside this, food-packaging material should fulfill at least three functional properties such as mechanical protection of food product during transport and storage, air and moisture barrier and protection against contamination by microorganisms. Thus, researchers are giving emphasis on the development of film, which would be biodegradable but also offers better oxygen barrier and water barrier properties. Though the concept of *hydrogel food packaging* is relatively new like green packaging or sustainable packaging, it does not yet come into the forefront for its use in practicality.

4.1.3 USE OF HYDROGELS FOR FOOD PACKAGING

Application of hydrogels from bio-based materials offers the potential for developing novel, biodegradable packaging applications, particularly for foods, that meet the ever-increasing demands for natural and environmentally compatible materials. Development of food packaging applications from biopolymers has lagged behind medical materials due to high cost, low strength, and poor water resistance (25).

Packaging can be divided into three broad categories: Primary packaging (in contact with the goods and taken home by consumers), secondary packaging (covers the larger packaging, i.e. boxes, used to carry quantities of primary packaged goods) and tertiary packaging (used to assist transport of large quantities of goods, i.e. wooden pallets and plastic wrapping). Generally, secondary and tertiary packaging materials are easy to collect and sort out for recycling. But in the case of primary packaging materials, they are largely mixed and contaminated, thus create problems in recycling or reuse of the materials [25]. Different materials are being used for packaging according to the nature of the material to be packed [26]. Glass and metals are also widely used for packaging as they are resistant to corrosion, and strong enough. But, their uses have been restricted to a certain limit considering the cost, weight and users inconvenience. However, among the packaging materials, polymers, specifically plastics are in high demand. They exhibit many worthwhile characteristics, such as transparency, softness, heat seal ability, and good strength to weight ratio. In addition, they are generally low-cost materials, show efficient mechanical properties such as tear and tensile strength, and they are good barrier to oxygen and heat [26, 29]. But, most of the plastic food packaging materials are practically non-degradable and also present the serious problem of environmental pollution.

As an alternative, hydrogels can offer new opportunities for design of efficient biopolymer packaging materials with desirable properties [30]. Hydrogels are frequently found in everyday products such as contact lens, capsules for oral ingestion, coatings, membranes, slabs, micro- and nanoparticles. Cast into films and dried, hydrogels now are also being tailored for biodegradable packaging materials for food, cosmetic and pharmaceutical products [31].

Dry hydrogels from biomacromolecules present a number of desirable properties for packaging films, particularly biodegradability and the possibility to incorporate cells, bioactive compounds and drugs. Furthermore, due to the chemical properties of functional groups along the molecule backbone, hydrogels can be developed as 'smart' tailored devices able to respond to specific external stimuli (e.g., pH and temperature) that act as triggers to modify over time the release rates of compounds loaded in them [32]. Therefore, stimuli-sensitive hydrogels differ from inert hydrogels in that they can "sense" changes in environmental properties and respond by increasing or decreasing their degree of swelling and thus their hydration state.

Formation of physical hydrogels from single biopolymers is a well-known phenomenon generating structure in foods [33], but single-polymer hydrogels tend to form weak films due to limited interchain interactions. In contrast, generating physical hydrogel films from combinations of polyelectrolyte biomolecules with opposite charges offers distinct advantages.

Hydrogels formed from starch and PVA, with or without addition of polycarboxylic acids, have demonstrated a high-water retention capacity [34]. This property can be used to incorporate into the hydrogel substances capable of acting as antimicrobials, thus limiting the growth of bacteria and fungi responsible for the deterioration of food. One of the most commonly used antimicrobial substances in new packaging applications is essential oil. These substances, extracted from different plant species generally contain a large amount of phenolic compounds and terpenes, which confer a high antimicrobial and antioxidant capacity. Among these oils, one of the most effective is cinnamon oil, which is the volatile liquid fraction obtained from the dried bark of the plant *Cinnamomum zeylanicum*. This hydrophobic mixture contains as main components cinnamaldehyde and eugenol [34]. In addition, it is important to note that the use of chitosan and fungicides is a key factor in the conservation of strawberries and raspberries [35], citrus [36], cut chestnuts [36]. Studies with different fruits and vegetables: blueberries, kiwis, cherry, squash, pineapple, apples, potato, celery and strawberries, in which different types of coatings were applied, the authors observed that chitosan reduced the decay rate at room temperature, and the caseinates of calcium and sodium alginate were good barrier to oxygen, delaying the maturation of the fruit and maintaining the firmness [32–37].

Given the multifunctionality of pectin and gelatin, various combinations of electrostatic interactions can be exploited to create mixed gelatin-pectin physical hydrogels as starting materials for packaging films. The final hydrogel characteristics will depend on the specific combinations of materials and conditions. Appropriate operative conditions for the fabrication of physical hydrogels must be determined for each gelatin-pectin combination because each system will behave in its own specific fashion; a single standard procedure cannot be established to cover all hydrogel variants [40, 41]. Table 4.1 is a summary of the different packages obtained when mixing protein with polysaccharides:

4.2 CONCLUSIONS

Hydrogels used as edible coatings made from biopolymers have numerous advantages over those synthesized from synthetic polymers, including being biodegradable, recyclable, can carry additives, possess good mechanical and barrier properties, besides improving the appearance of food and protecting its properties during storage and handling. They also maintain or improve sensory characteristics and textures in food add nutritional value to the product, specifically those produced from protein.

TABLE 4.1 Mixtures of Biopolymers Used in Food Packaging [32]

Mixture	Form	Application	Advantage/disadvantage
Chitosan	Hydrogel	Food packaging of fruits (apple, pear and grape)	Biodegradable and biocompatible
Starch/PVA	Hydrogel with substances antimicrobials	Packaging strawberries and raspberries	Biodegradable
Protein/mesquite gum	Films	Food wrapping	Flexibility/yellowish color
Beta-lactoglobulin/pectin	Nano-particles	Vehicles for $u-3$ polyunsaturated fatty acids	Colloidal stability, transparent dispersions
Chitosan/gelatin	Membranes	Biomedical (tissue engineering)	Bioactivity/dissolution of gelatin from membranes
Starch/cellulose fibers	Films	Food packaging	High tensile strength/poor optical properties
Chitosan/pectin	Coatings	Plastic packaging for post-harvest crop protection	Antimicrobial activity/high coating thickness
Gelatin/sodium alginate	Films	Edible casings	Oxygen barrier/shrinkage over time
Gelatin/maltodextrin	Gels	Release of active compounds	Phase separation
Chitosan/gelatin	Coatings	Antimicrobial packaging	Antimicrobial properties/yellowness
Gelatin/pectin	Films	Food coverings	Casting high transparency
Methylcellulose/whey protein	Films casting water vapor	Moisture sensitive food products	Barrier/potential phase separation
Gelatin/can bagasse	Films	Self-fertilizing mulching biopolymers	Thermal stability/rigidity, dark color
Whey protein/pectin	Complexes structure	Formation and stabilization of food systems	Tailored properties/turbidity

KEYWORDS

- biodegradable
- edible coatings
- hydrogels
- synthetic polymers

REFERENCES

1. Perdones, A., Vargas, M., Atarés, L., & Chiralt, A., (2014). Physical, antioxidant and antimicrobial properties of chitosan-cinnamon leaf oil films as affected by oleic acid. *Food Hydrocolloids, 36*, 256–264.
2. Almenar, E., Catalá, R., Hernández, P., & Gavara R., (2009). Optimization of an active package for wild strawberries based on the release of 2-nonanone. *Food Science and Technology, 42*(2), 587–593.
3. Soliva, R. C., & Martín, O., (2003). New advances in extending the shelf life of fresh-cut fruits: A review. *Trends in Food Science and Technology, 14*(9), 341–353.
4. Ahvenainen, R., (2000). Ready-to-use fruit and vegetables. *Technical Manual, 376*, 1–10
5. Villada, H. S., Acosta, H. A., & Velasco, R. J., (2007). Biopolymers naturals used in biodegradable packaging. *Temas Agrarios., 12*(2), 5–13.
6. Tharanathan, R. N., (2003). Biodegradable films and composite coatings: Past, present and future. *Trends in Food Science and Technology, 14*, 71–78
7. Park, K., Chen, J., & Park, H., (2001). Hydrogel composites and superporous hydrogel composites having fast swelling, high mechanical strength, & superabsorbent properties. *US Patent no 6271278.*
8. Krochta, J. M., & De Mulder, C., (1997). Edible and biodegradable polymer films: Challenges and opportunities. *Food Technology, 51*, 61–74
9. Fleming, M. E., & Smith, R. E., (2003). Volatile organic compounds in foods: A five year study. *Journal of Agricultural and Food Chemistry, 51*(27), 8120–8127.
10. Liang, H., Yuan, Q., Vriesekoop, F., & Lv, F., (2012). Effects of cyclodextrins on the antimicrobial activity of plant-derived essential oil compounds. *Food Chemistry, 135*(3), 1020–1027.
11. Bejarano, L., Rojas, B., Prin, J. L., Mohsin, M., García, A., Mostue, M. B., et al., (2008). Katime I. Synthesis and study of hydrogels obtained from acrylamide, poly (acrylic acid) and maleic acid as potential remedies for metallic contaminants in wastewater. *Rev. Iberoam. Polim., 9*(3), 307–312.
12. Ramírez, A., Benitez, D., Guzman, P., & Rojas, B., (2011). Interacciones de hidrogeles de poli(acrilamida–co–ácido itacónico) estudiadas en soluciones de Ca(NO$_3$). *Rev. Iberoam. Polim., 12(6), 308–316.*

13. Peppas, N. A., & Stauffer, S. R., (1991). Reinforced uncrosslinked poly (vinyl alcohol) gels produced by cyclic freezing-thawing processes: A short review. *Journal of Controlled Release, 16*(3), 305–310.
14. Lowman, A. M., & Peppas, N. A., (1999). Hydroges, In: Mathiowitz, E., (ed.), *Encyclopedia of Controlled Drug Delivery* (pp. 397–418). Wiley, New York.
15. Gao, Y., Xu, S., Wu, R., Wang, J., & Wei, J., (2008). Preparation and characteristic of electric stimuli responsive hydrogel composed of polyvinyl alcohol/poly(sodium maleate -co-sodium acrylate). *Journal of Applied Polymer Science, 107*, 391–395.
16. Hennink, W. E., & Nostrum, C. F., (2002). Novel cross-linking methods to design hydrogels. *Advanced Drug Delivery Reviews, 54*(1), 13–36.
17. Çiçek, H., & Tuncel, A., (1988). Immobilization of α-chymotrypsin in thermally reversible isopropylacrylamide-hydroxyethylmethacrylate copolymer gel. *Journal of Polymer Science Part A: Polymer Chemistry, 36*(4), 543–552.
18. Bettini, R., Colombo, P., & Peppas, N. A., (1995). Solubility effects on drug transport through pH-sensitive, swelling-controlled release systems: Transport of the ophylline and metoclopramide monohydrochloride. *Journal of Controlled Release, 37*(1–2), 105–111.
19. Carrillo, M., Vivas, M., Jiménez, L., Hernández, L., Ramirez, M., & Katime, I., (2009). Síntesis de hidrogeles de poli(ácido itacónico-cometacrilato de metilo). *Rev. Iberoam. Polím., 10*(4), 188–195.
20. Rojas, B., Ramírez, M., Aguilera, R., García, A., Prin, J. L., Lias, J., et al., (2007). Hydrogels obtained from acrylamide, maleic acid acrylic acid and octylmonoitaconate: Synthesis, absorbent capacity and pH variations in copper sulfate solu tions. *Rev. Téc. Ing. Univ., 30*(1), 74–84.
21. Hoffman, A. S., (2002). Hydrogels for biomedical applications. *Advanced Drug Delivery Reviews, 43*, 3–12.
22. Hoare, T. R., & Kohane, D. S., (2008). Hydrogels in drug delivery: Progress and challenges. *Polymer, 49*, 1993–2007.
23. Jo, C., Kang, H., Lee, N. Y., Kwon, J. H., & Byun, M. W., (2005). Pectin and gelatin-based film: Effect of gamma irradiation on the mechanical properties and biodegradation. *Radiation Physics and Chemistry, 72*, 745–750.
24. Terao, K., et al., (2003). Reagent-free cross-linking of aqueous gelatin: Manufacture and characteristics of gelatin gels irradiated with gamma ray and electron beam. *Journal of Biomaterials Science-Polymer, 14*, 1197–1208.
25. Liu, L. S., Fishman, M. L., Hicks, K. B., & Kende, M., (2005). Interaction of various pectin formulations with porcine colonic tissues. *Biomaterials, 26*, 5907–5916.
26. Peppas, N. A., Huang, Y., Torres-Lugo, M., Ward, J. H., & Zhang, J., (2000). Physicochemical foundations and structural design of hydrogels in medicine and biology. *Annual Review of Biomedical Engineering, 2*, 9–29.
27. Xiao, C., & Yang, M., (2006). Controlled preparation of physical cross-linked starch-g-PVA hydrogel. *Carbohydrate Polymers, 64*(1), 37–40.
28. Gao, Y., & Xiao, C., (2005). Preparation and characterization of starch-g-PVA/nano-hydroxyapatite complex hydrogel. *Materials Science, 20*, 58–59.

29. Burt, S., (2004). Essential oils: Their antibacterial properties and potential applications in foods – A review. *International Journal of Food Microbiology, 94*(3), 223–253.
30. Giménez, B., Gómez, M. C., & Montero, P., (2005). The role of salt washing of fish skins in chemical and rheological properties of gelatin extracted. *Food Hydrocolloids, 19*(6), 951–957.
31. Choi, S. S.,& Regenstein, J. M., (2000). Physicochemical and sensory characteristics of fish gelatin. *Journal of Food Science, 65*(2), 194–199.
32. Farris, S., Introzzi, L., & Piergiovanni, L., (2009). Evaluation of a biocoating as a solution to improve barrier, friction and optical properties of plastic films. *Packaging Technology and Science, 22,* 69–83.
33. Gomez, J., Montero, P., Fernández, F., & Gomez, M. C., (2009). Physico-chemical and film forming properties of bovine-hide and tuna-skin gelatin: A comparative study. *Journal of Food Engineering, 90*(4), 480–486.
34. Mangiacapra, P., Gorrasi, G., Sorrentino, A., & Vittoria, V., (2006). Biodegradable nanocomposites obtained by ball milling of pectin and montmorillonites. *Carbohydrate Polymers, 64,* 516–523.
35. Qiu, Y., & Park, K., (2001). Environment-sensitive hydrogels for drug delivery. *Advanced Drug Delivery Reviews, 53,* 321–339.
36. Dickinson, E., (2006). Colloid science of mixed ingredients. *Soft Matter, 2,* 642–652.
37. Durango A. M., Soares N. F., & Arteaga M. R., (2011). Filmes y revestimientos comestibles como empaques activos biodegradables en la conservación de alimentos. *Biotecnología en el Sector Agropecuario y Agroindustrial, 9*(1), 122–128.
38. Gao,Y., & Xiao, C., (2005). Preparation and characterization of starch-g-PVA/nanohydroxyapatite complex hydrogel. *Materials Science, 20,* 58–59.
39. Chen, Z., Mingzhu, L., & Songmei, M., (2005). Synthesis and modification of salt-resistant superabsorbent polymers. *React. and Function. Polym., 62*(1), 85–92.
40. Li, B., Kennedy, J. F., Jiang, Q. G., & Xie, B. J., (2006). Quick dissolvable, edible and heatsealable blend films based on konjac glucomannan e gelatin. *Food Research International, 39,* 544–549.
41. Avena, R. J., Olsen, C. W., Olson, D. A., Chiou, B., Yee, E., Bechtel, P. J., et al., (2006). Water vapor permeability of mammalian and fish gelatin films. *Journal of Food Science, 71,* 202–207.

CHAPTER 5

COFFEE PULP AS POTENTIAL SOURCE OF PHENOLIC BIOACTIVE COMPOUNDS

LUIS V. RODRÍGUEZ-DURÁN,[1] ERNESTO FAVELA-TORRES,[1] CRISTÓBAL N. AGUILAR,[2] and GERARDO SAUCEDO-CASTAÑEDA[1]

[1] *Department of Biotechnology, Autonomous Metropolitan University, Campus Iztapalapa, Av. San Rafael Atlixco 186, Zip Code 09340, Iztapalapa, Mexico City, Mexico, Tel.: +52 55-58 04-4600, fax: +52 55-580-4499, E-mail: saucedo@xanum.uam.mx*

[2] *Department of Food Science and Technology, School of Chemistry, Universidad Autónoma de Coahuila, Blvd. V. Carranza and González Lobo s/n, Zip Code 25280, Saltillo, Coahuila, Mexico*

ABSTRACT

Coffee pulp (CP) is the main solid by-product of the wet processing of coffee cherries and it represents about 40% of the cherry's fresh weight. More than 400,000 tons of coffee pulp is produced in Mexico per year. Traditional applications utilize only a fraction of the available material. Therefore, it is necessary to find alternative uses for this by-product. One of the most promising applications for the use of coffee pulp consists of the extraction purification and biotransformation of valuable phenolic compounds. Through this process, powerful natural antioxidants, with diverse industrial applications can be obtained. Furthermore, phenolics free CP could be used in some other application, such as bio-fertilizer, animal feed, among others.

Several classes of polyphenols have been identified in coffee pulp such as flavanols, flavonols, anthocyanidins, hydroxycinnamic acids and their derivatives. However, the amount and distribution of phenolics in CP may differ radically from one study to another. This variability could be due to the origin of coffee samples (e.g., cultivar, place of origin, culture conditions, ripening stage, etc.), the post-harvest management (methods of pulping, drying, storage, etc.), or even to the analytical methods used.

In this chapter, we will review the available information on the content of phenolic compounds in CP, as well as the methods used for its extraction, recovery, isolation, and purification. The biotransformation and enzymatic modification of these phenolic compounds will also be discussed.

5.1 INTRODUCTION

Coffee is one of the most consumed beverages in the world. It is an infusion prepared from the roasted and ground beans of *Coffea* sp. plants, mainly from the *C. arabica* and *C. canephora* species. Coffee growing and processing is an agro-industry of great economic and social importance in many developing countries. Currently, coffee is produced in more than 50 countries and the annual production is about 9 millions of tons [1]. On the other hand, the coffee agro-industry is also an important source of waste and pollution. It is estimated that this industry generates only in Mexico more than half a million tons of solid by-products per year [2].

Coffee is internationally traded as green coffee (the coffee bean covered or not with the silverskin). To obtain green coffee, coffee fruits must be processed by either dry or wet method. In the dry process, ripe coffee fruits, also called coffee cherries, are sun-dried and mechanically hulled; the dried husks (skin, pulp, mucilage, and parchment) are removed together with part of the silverskin. In the wet process, the skin and most of the pulp are mechanically removed by pressing the fresh fruits in water through a screen in a pulper. Pulp remnants and the mucilage layer are removed either mechanically or by natural fermentation. The pulped coffee is dried and the parchment removed with a hulling machine [3].

Coffee pulp (CP) is the main solid by-product from the wet processing of coffee cherries and it represents about 40% of the fresh weight of coffee fruits [4]. Traditional applications of CP include their use for animal feed [5], the cultivation of edible mushrooms [6] and the production of compost [7] and vermicompost [4]. More than one million tons of coffee cherries and 400,000 tons of CP are produced in Mexico per year [8]. However, only a part of the available CP is used by these applications. The remaining CP can cause serious environmental problems if not properly disposed of. Therefore, it is necessary to find alternative uses for this by-product.

One of the most promising potential applications of CP is the extraction of valuable phenolic compounds contained in this by-product. This process can obtain compounds with important biological activities and industrial applications. In addition, CP free of phenolics can be used for some of the traditional applications mentioned above.

In this chapter, we will review the available information on the content of phenolic compounds in CP, as well as the methods used for its extraction, recovery, isolation, and purification. The biotransformation and enzymatic modification of coffee phenolic compounds will also be discussed.

5.2 COFFEE PULP

CP consists of the outer skin or pericarp of the fruit, and most of the mesocarp [3]. The presence of toxic and anti-nutritional compounds limits its use in various applications, such as in animal feed [9]. In addition, fresh pulp has a high moisture content (about 77%), which makes it difficult to handle and transport [10].

In 2015, 1.026 million tons of cherry coffee were produced in Mexico and 86% of that production was processed through the wet method, generating more than 400,000 tons of fresh CP [8, 11]. This represents more than 100,000 tons on a dry basis. Therefore, several researchers have developed processes for the use of CP (Table 5.1). However, few of these techniques are applied in Mexico, so much of the CP is dumped in large open piles in gorges or near to rivers, causing both water and soil pollution [12].

5.2.1 APPLICATIONS OF CP

CP contains important quantities of nutrients that could be used in animal feed. However, its high content of toxic and anti-nutritional compounds limits its use for this purpose. It has been found that it can be used as animal feed for dairy cows up to 20% of CP without a negative effect on their growth. However, for other animals, such as roosters, the limit is as low as 5% by weight [5]. An alternative for the use of CP is the detoxification prior to its use as animal feed. This detoxification can be carried out by silage and or by solid state fermentation (SSF) with filamentous fungi [26].

TABLE 5.1 Applications and Potential Uses of CP

Product	Reference
Animal feed	[13]
Organic fertilizer	[4]
Solid fuel	[14]
Bioethanol	[15]
Biogas	[16]
Edible mushrooms	[6]
Aroma compounds	[17]
Phenolic compounds	[18]
Enzymes (α-amylase, xylanase, pectinase, laccase, tannase, feruloyl esterase, chlorogenate hydrolase, etc.)	[19–25]

CP can also be used as an organic fertilizer, mainly to fertilize coffee trees. But, the use of fresh CP as fertilizer has a drawback: Its handling is complicated due to high humidity. Therefore, it has been proposed to dry the pulp for use as fertilizer. However, to match the chemical composition of one kg of inorganic fertilizer, it is required 5 to 10 kg of dry CP [27]. An alternative to improve the properties as a biofertilizer, CP can be subjected to composting or vermicomposting process [4, 7]. Vermicomposting is a technique based on the fact that certain species of worms such as *Eisenia andrei*, *E. fetida*, and *Perionyx excavates* consume organic waste very rapidly and convert them into material rich in available nutrients to plants. At the end of the process, the quantity of carbon is significantly reduced, so the C/N ratio of CP is reduced from 30:1 to 12:1 in the finished fertilizer. Other elements in CP maintain their total amount and, therefore, they are concentrated in the final product [4].

Another of the conventional applications of CP is its use as fuel. Traditionally CP has been used for direct burning in the drying of the coffee beans [28]. This allows saving energy during the mechanical drying of coffee. However, not all coffee processing facilities use mechanical dryers. The application of CP as fuel is very limited due to its low energy density. An alternative to the use of this by-product is the production of compressed solid fuels, such as pellets and briquettes [14].

CP has also been used to produce gaseous and liquid fuels. It has been reported that 133 m^3 of biogas can be produced from one ton of CP by anaerobic digestion. This is equivalent to the energy content of 100 L of oil [29]. The main drawback of this process is the low economic value of biogas well as the cost of transporting the raw materials and gas produced. The use of CP to produce bioethanol has been investigated [15, 30, 31]. The production of ethanol from CP has the drawback of requiring a pretreatment and/or hydrolysis steps to release the fermentable sugars. Good yields (from 78 to 87% of maximum theoretical yield) have been obtained through this process, but low concentrations of ethanol (from 12 to 16 g/L) were obtained due to the low concentration of fermentable sugars in coffee extracts [30, 31].

CP is a good substrate for the production of edible fungi such as *Pleurotus*, *Lentinula* and *Auricularia* [6]. In Mexico, rural cultivation of mushrooms using coffee residues as a substrate began in 1989 in an indigenous community in Puebla and has spread to other regions of the country due to the social, economic and ecological benefits for rural development [32]. The cultivation of mushrooms at the rural level has been successful and sustainable [6]. However, a number of technical and economic problems have prevented bring this process to an industrial level [6].

The production of aromas using CP as substrate has been studied. Medeiros et al. [17] studied the ability of two strains of *Ceratocytis fimbriata* to produce fruity aromas in SSF using CP as solid support and substrate. Eleven volatile compounds were detected during fermentation, of which ethyl acetate and ethanol were the most abundant.

CP has been used as a substrate to produce different enzymes in SSF. The composition of CP for the production of enzymes acting on phenolic compounds, such as tannase [25], laccase [22, 33], feruloyl esterase [23], and chlorogenate hydrolase [24]. This by-product has also been used for the production of carbohydrate degrading enzymes, such as pectinases [21, 34], xylanase [20] and α-amylase [19].

An application that has recently attracted attention is the extraction of phenolic compounds present in CP. This can be carried out by chemical or biological processes, such as the application of enzymes or by biotransformation by microorganisms by SSF [35–37].

5.2.2 CHEMICAL COMPOSITION OF CP

Fresh CP has high moisture content (greater than 70%), which makes it difficult to handle. For most applications, CP must be dried to prevent microbial growth during storage. The chemical composition of the dried CP is shown in Table 5.2.

CP has a high content of carbohydrates (66.5–84.4%) and protein (8–15%) so it could have a high nutritional value. But it also has significant amounts of anti-nutritional compounds, such as tannins (0.7–4.6%) and caffeine (0.7–1.8%) [38–43].

In addition to the cell wall components (cellulose, hemicellulose, and lignin), the carbohydrate fraction of CP contains: Reducing sugars (12.40%), non-reducing sugars (2.02%) and pectic substances (6.52%) [38].

5.2.3 PHENOLIC COMPOUNDS IN CP

One of the most attractive potential applications for the use of CP is the extraction of phenolic compounds, due to its high commercial value. For

TABLE 5.2 Chemical Composition of CP

Component	Content (% w/w, dry basis)	Reference
Proximate composition		
Crude fat	1.3–3.8	[38–43]
Crude fiber	16.2–35.2	[38–42]
Protein	8.0–15.0	[38–43]
Ash	5.0–9.5	[38–43]
Nitrogen free extract	42.7–63.5	[38–42]
Carbohydrate*	66.5–84.4	[38–42]
Cell wall components		
Cellulose	10.5–28.6	[38, 40, 41, 43]
Hemicellulose	1.0–15.5	[38, 40, 41]
Lignin	12.2–22.8	[38, 40, 41]
Other organic components		
Tannin	0.7–8.6	[38, 40, 41, 43]
Caffeine	0.7–1.8	[38, 40, 41, 43]

*Carbohydrate = Crude fiber + Nitrogen free extract.

this reason, the phenolic compounds of CP have been extensively studied for more than 20 years. Different phenolic compounds have been identified and quantified in this by-product, such as flavanols, flavonols, anthocyanidins, proanthocyanidins, hydroxybenzoic acids, hydroxycinnamic acids and their derivatives (Table 5.3).

TABLE 5.3 Content of Phenolic Compounds in CP

Compound	Content (g/kg, dry basis)*	References
Flavanols		
Catechin	0.06–0.63	[44]
Epicatechin	1.53–4.36	[44]
Total flavanols	0.60–4.94	[35, 44]
Flavonols		
Rutin	0.09–0.70	[44]
Total flavonols	0.09–0.70	[35, 44]
Anthocyanidins	0.040–0.050	[35]
Proanthocyanidins		
Prodelphinidins	6.80–18.6	[48]
Procyanidins	1.20–8.50	[48]
Total proanthocyanidins	8.60–27.1	[35, 48, 49]
Hydroxybenzoic acids		
Protocatechuic acid	0.09–4.70	[44, 45]
Total hydroxybenzoic acids	0.09–4.70	[44, 45]
Hydroxycinnamic acids and derivatives		
Caffeic acid	0.09–2.66	[18, 46, 47, 50]
Ferulic acid	0.05–0.26	[18, 44, 46, 47, 50]
p-coumaric acid	0.08–0.16	[18, 46, 47, 50]
5-caffeoyl quinic acid (chlorogenic acid)	2.38–12.80	[18, 35, 44–46, 50]
4-caffeoyl quinic acid	0.30–0.30	[45]
3-caffeoyl quinic acid	0.20–0.20	[45]
3,5-dicaffeoyl quinic acid	0.20–1.13	[44, 45]
3,4- dicaffeoyl quinic acid	0.90–4.45	[44, 45]
4,5-dicaffeoyl quinic acid	0.00–1.22	[44, 45]
5-feruloyl quinic acid	0.00–0.20	[45]
p-cumaroyl quinic acid	0.08–0.60	[35]
Total hydroxycinnamic acids and derivatives	2.98–15.98	[18, 35, 44–47, 50]

5.2.3.1 FLAVONOIDS

Flavonoids are polyphenolic compounds of 15 carbon atoms formed by two aromatic rings connected by a heterocyclic pyran ring [51]. These compounds are widely distributed in the plant kingdom and are present in high concentrations in the epidermis of the leaves and in the fruit skin [52]. The main subclasses of flavonoids are flavones, flavonols, flavanols, isoflavones, flavanones, and anthocyanidins. Figure 5.1 shows the generic structures of the major subclasses of flavonoids. These generic structures can have numerous substitutions. The most common are hydroxyl and methyl groups and glucosides. These molecules can be joined to give oligomeric or polymeric proanthocyanidins, also known as condensed tannins [52].

Different classes of flavonoids and derivative compounds have been identified in CP. Ramírez [53] analyzed the phenolic compounds present in the methanolic extracts of CP. He tentatively identified compounds belonging to the groups of catechins, leucoanthocyanidins, anthocyanins and glycoside derivatives of flavonols by paper chromatography. In subsequent studies, the flavonoids catechin, epicatechin, and rutin were identified and quantified by HPLC [44].

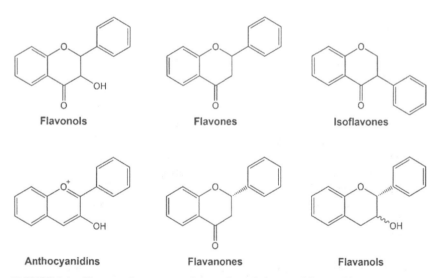

FIGURE 5.1 The generic structure of the major subclasses of flavonoids.

More recently, Ramirez-Coronel et al. [35] analyzed the phenolic compounds of CP by direct thiolysis followed by HPLC. They found flavanols in the form of monomers and proanthocyanidins. The constitutive units of the proanthocyanidins were mainly epicatechin (more than 90%). They also found small amounts of flavonols and anthocyanidins (0.4–0.6 g/kg, dry basis).

5.2.3.2 PHENOLIC ACIDS

Phenolic acids are compounds having an aromatic ring with at least one hydroxyl and a carboxyl group (Figure 5.2). Natural phenolic acids are classified according to their structure in hydroxybenzoic and hydroxycinnamic acids [54]. The most abundant phenolic acids in CP are hydroxycinnamic acids and their derivatives. The only hydroxybenzoic acid identified in CP is the protocatechuic acid, which is in a concentration range of 0.09 to 4.70 mg/kg on a dry basis [44, 45].

The amount and distribution of hydroxycinnamic acids in CP depends on the origin of the coffee fruits (cultivar, place of origin, culture

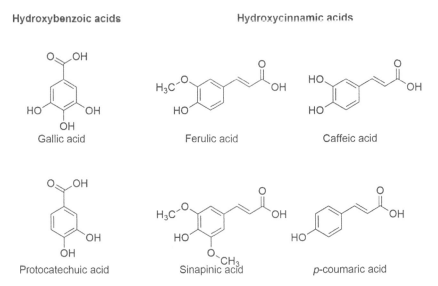

FIGURE 5.2 Structure of the main phenolic acids.

conditions, maturity, etc.) and the post-harvest management (methods of pulping, drying, storage, etc.). In addition, the use of different analytical methods can drastically change the results obtained [50].

The hydroxycinnamic acids can be in the form of free acids, soluble esters and insoluble esters attached to the cell wall components. Several hydroxycinnamic acids have been identified and quantified, such as caffeic, ferulic and *p*-coumaric acid, as well as their soluble esters: 3-caffeoyl quinic acid, 4-caffeoyl quinic acid, 5-caffeoyl quinic acid, 3,5-dicaffeoyl quinic acid, 3,4-dicaffeoyl quinic acid, 4,5-dicaffeoyl quinic acid, 5-feruloyl quinic acid and *p*-coumaroyl quinic acid. The 3-caffeoyl quinic acid, also known as chlorogenic acid is the most abundant phenolic acid in CP. It is found in the range of 1 to 25 mg/g of CP (dry basis) and it represents about 50–98% of total hydroxycinnamic acids and derivatives [18, 35, 44–47, 50].

Several *in vitro*, *in vivo* and epidemiological studies have attributed potential benefits of consumption of chlorogenic acid to human health. These potential benefits are based on their antioxidant, antiviral, hypoglycemic and hepatoprotective properties [55]. Therefore, chlorogenic acid has potential applications in the chemical, pharmaceutical, and food industry. Currently, commercial chlorogenic acid is obtained as a mixture of isomers extracted from green coffee beans and some medicinal plants, such as the honeysuckle (*Lonicera japonica*) and *Eucommia ulmoides* [56]. The commercial value of chlorogenic acid ranges from 36 to 65,000 USD per kg, depending on the degree of purity (10–95%) (Xi'an Tianxingjian Natural Bio-Products Co., Ltd., China; Sigma-Aldrich Co. LLC, USA). Therefore, it may be attractive to obtain the chlorogenic acid from low-cost sources, such as CP.

5.2.3.3 TANNINS

The term tannin groups a number of phenolic compounds of variable structure, but with some characteristics in common. These compounds are water-soluble and have the ability to complex and precipitate proteins and other macromolecules [57]. Tannins are divided into two great groups, namely condensed and hydrolyzable tannins. Hydrolyzable tannins are hydroxybenzoic acid derivatives containing a central core of glucose (or another polyol) esterified with gallic or hexahydroxydiphenic acids.

Condensed tannins, also called proanthocyanidins, are oligomers or polymers of flavonoids [58]. CP tannins are mainly from the condensed group.

Several authors have analyzed the content of tannin in CP by spectrophotometric methods, such as the Folin-Ciocalteu or the Porter's reagent method. These techniques are fast and simple, but unspecific and, therefore, susceptible to multiple interferences. García et al. [59] estimated the content of total phenolics and condensed tannins of sun-dried CP in 42.5 and 25.6 mg/g (dry basis). Clifford and Ramírez-Martínez [48] reported a content of soluble condensed tannins of 9.5 and 27.1 mg/g (dry basis) for *C. arabica* and *C. canephora* pulp, as well as the absence of hydrolyzable tannins.

De Colmenares et al. [49] studied the content of condensed tannins in fresh and air dried CP of three *C. arabica* cultivars. They found a higher content of condensed tannins in the yellow variety (Yellow Catuai) with respect to two red varieties (Red Catuai and Red Bourbon). They also observed a higher concentration of condensed tannins during the dry CP compared to the fresh pulp, as well as an increase in the oligomeric proanthocyanidins together with a decrease of the dimeric proanthocyanidins in the dry CP. This suggests a transformation of the dimers to higher molecular weight oligomers during the drying of CP.

5.3 EXPLOITATION OF CP POLYPHENOLS

The use of CP for obtaining valuable phenolic compounds can be an economically feasible alternative for the valorization of this by-product, which is largely underutilized in our country. In addition, the utilization of CP would contribute to reducing the pollution generated by the coffee industry. The exploitation of the phenolic compounds of CP requires several operations such as extraction, isolation, and purification. In addition, these phenolic compounds can be subjected to chemical or biological treatments to obtain different compounds with increased biological activity or higher commercial value.

5.3.1 EXTRACTION

The extraction of phenolic compounds from CP for analytical purposes have been carried out mainly by conventional solid-liquid extractions.

The methodologies used for the extraction of flavonoids and condensed tannins from CP differ considerably from those used for the extraction of phenolic acids and hydrolyzable tannins.

Ramírez-Coronel et al. [35] studied the efficiency of extraction of phenolic compounds with different solvents. They performed a series of successive extractions with hexane, acidified methanol, and acidified aqueous acetone. The treatment with methanol solubilized most of the monomeric flavanols (63 to 75%) as well as hydroxycinnamic acids (60 to 66%). The majority of proanthocyanidins were solubilized in the acetone extract (55 to 63.5%).

Most of the studies on the phenolic acids of CP are limited extracts obtained with aqueous methanol solutions. However, significant amounts of phenolic compounds may remain in the solid after such extractions. Pérez-Morales et al. [23] optimized the extraction of hydroxycinnamic acids from CP with aqueous methanol. They evaluated the concentration of the solvent, the temperature and the extraction time. Under the best conditions (80% methanol, 56°C and 34 minutes) 100.1 mg of hydroxycinnamic acids and derivatives per gram of CP were extracted. This represents about 3% of total hydroxycinnamic acids and derivatives in CP as determined by methanol extraction followed by alkaline and acid hydrolysis.

There are improved solid-liquid extraction techniques such as microwave-assisted extraction and ultrasonic assisted extraction. These techniques allow a high recovery of the compounds of interest in a short extraction time and with reduced solvent consumption. Tobón Arroyave [60] compared ultrasound assisted extraction and conventional solid-liquid extraction for the recovery of phenolic compounds of CP. Under the best conditions (50% ethanol, 45 kHz, 300 W and 45 min of extraction), 63.3 mg of gallic acid equivalents per gram of CP were extracted by ultrasound-assisted extraction. This represents about 90% of that obtained by conventional extraction with methanol, but at infraction of the time required and by using a less toxic solvent. Rodríguez-Durán et al. [61] used microwave-assisted extraction for the recovery of hydroxycinnamic acids from CP. They evaluated the effect of the solvent, the solvent concentration and the treatment time on the extraction of hydroxycinnamic acids. Under the best conditions (50% ethanol, 400 W, and 2 min), they recovered 100% of the hydroxycinnamic acids extracted from CP by successive extractions with hexane and methanol (4.19 ± 0.2 mg/g). Moreover, microwave-assisted extraction was carried out in a single step, in a short time (2 min) and using an environmentally friendly solvent.

Solid liquid extraction (conventional or improved) can be used to extract free phenolic acids and soluble derivatives, such as chlorogenic acid and its isomers. However, there is a fraction of phenolic compounds that are covalently bound to the polymers of the cell wall and can only be released by hydrolysis. The extraction of covalently bound phenolic acids is generally carried out by alkaline hydrolysis. Torres-Mancera et al. [18] reported the composition of free and covalently bound hydroxycinnamic acids in CP. They found that most phenolic acids (83.7%) are in the form of insoluble esters.

Alkaline hydrolysis is commonly used for the extraction of covalently bound phenolic acids. However, it is known that this process led to an important loss of phenolic compounds, particularly dihydroxy-derivatives such as caffeic acid [62]. For example, Krygier et al. [63] reported 66.7% caffeic acid lost when treated with 2 M NaOH for 4 h even under a nitrogen atmosphere. In order to prevent loss of phenolic acids, Nardini et al. [62] proposed the addition of EDTA and ascorbic during alkaline hydrolysis. Under their assay conditions (2 M NaOH containing 10 mM EDTA and 1% ascorbic acid at 30°C for 30 min) it was achieved the complete hydrolysis of chlorogenic acid without any detectable loss of caffeic acid.

Rodriguez-Durán et al. [50] evaluated the effect the alkaline hydrolysis conditions for recovery of the hydroxycinnamic acids from CP in the presence of EDTA and ascorbic acid. They observed an increase in the amount of caffeic acid followed by a decrease of this compound, probably due to the oxidation of phenolic acids. Maximal recovery of caffeic acid was obtained at using 2 M NaOH and incubating for 2 hours at 40°C. Trials with chlorogenic acid as internal standard showed 90% of caffeic acid recovery in the proposed conditions.

The enzymatic extraction of bound phenolic compounds may be a viable alternative to traditional chemical hydrolysis processes. For example, Benoit et al. [47] used a type B feruloyl esterase for the extraction of hydroxycinnamic acids from CP. With this treatment, caffeic acid was quantitatively extracted as well as the 73% of the *p*-coumaric acid and 45% of the ferulic acid.

Nieter et al. [64] utilized two recombinant type D feruloyl esterases from *Schizophyllum commune* (ScFaeD1 and ScFaeD2) for the release of hydroxycinnamic acids from CP. Significant amounts of caffeic (64–70%), ferulic (83–90%) and *p*-coumaric acids (100%) were released after an overnight incubation of CP with either ScFaeD1 or ScFaeD2.

Torres-Mancera et al. [18] used a crude extract produced by *Rhizomucor pusillus* in SSF and a commercial pectinase for the extraction of hydroxycinnamic acids from CP. Using this combination, they extracted 54.4% of the chlorogenic acid, 19.8% of the ferulic acid, 7.2% of the *p*-coumaric acid and 2.3% of the caffeic acid esterified in CP. In subsequent studies, using a combination of enzymatic extracts produced by *Aspergillus tamarii*, *Rhizomucor pusillus* and *Trametes* sp. 36.1% and 33% of the esterified chlorogenic and caffeic acid, respectively were extracted [65].

Ramírez-Velasco et al. [66] utilized an *Aspergillus ochraceus* enzymatic extract with chlorogenate esterase activity produced in SSF for the extraction of phenolic acids from CP. Under the selected conditions (40°C and pH 7) concentrated enzyme extract released more than 90% of the esterified caffeic acid contained in CP, in less than 20 min.

Another alternative for the extraction of phenolic compounds from CP is the SSF. In this process, the enzymes produced by the microorganisms can break the covalent bonds between the phenolic acids and the polysaccharides of the cell wall, facilitating their extraction. For example, Arellano-González et al. [37] studied the SSF of CP by *Aspergillus tamarii*. They found no significant differences in the total content of polyphenols in fermented and non-fermented CP, but there was an increase in the free phenolic acids (from 122 to 138 mg/g of CP) after fermentation and a decrease of the covalently linked polyphenols (from 213 to 185 mg/g of CP).

Bhoite et al. [67] studied the release of gallic acid from CP by *Penicillium verrucosum* in a SSF process. Under the optimal conditions (pH 3.32, moisture 58.41% and 96 h of fermentation at 30°C), 162.76 µg/g of CP were released.

Palomino-García [68] evaluated the effect of physicochemical parameters (moisture, pH, and temperature) on the polyphenol content of CP in a SSF process with *Penicillium purpurogenum*. In the best conditions, 170.3 mg of gallic acid equivalents per g of dry CP were released. The most abundant phenolic compounds found in this extract were chlorogenic acid, caffeic acid, and rutin with a concentration of 22.83, 4.29 and 1.95 mg/g of CP, respectively.

5.3.2 ISOLATION AND PURIFICATION

The extracts obtained from CP by the aforementioned methods consist of diluted solutions of polyphenols with different impurities. Depending on the application, extracts rich in phenolic compounds require some concentration and purity. The procedure for isolation and purification of polyphenols depends on the target compounds and the impurities in the crude extract.

The proanthocyanidins of CP can be isolated by means of liquid-liquid extractions and column chromatography. De Colmenares et al. [69] isolated two proanthocyanidin-containing preparations from the pulp of *C. canephora* cherries. The CP was extracted with aqueous acetone. Then, acetone was evaporated under reduced pressure and aqueous extracts were extracted with petroleum ether to remove lipids and fat-soluble pigments, with ethyl acetate to solubilize simple phenols, and with dichloromethane to eliminate the caffeine. The aqueous extract was applied to a Sephadex-LH 20 column, washed with water and ethanol and eluted with aqueous acetone (80% v/v). They isolated two proanthocyanidin-rich fractions with a slightly different composition, as can be inferred from IR and ^{13}C NMR spectra.

Murthy et al. [70] extracted and purified anthocyanins from CP. Anthocyanins were extracted with acidified aqueous methanol, the solvent was evaporated, concentrated in C18 sep-pak cartridges, loaded into a Sephadex-LH 20 column and eluted with a mixture of methanol/water/trifluoroacetic acid (20:79.5:0.5). The anthocyanins from CP yielded 0.25 mg of monomeric anthocyanins per g of CP. The purified anthocyanins were identified as cyanidin-3-rutinoside and cyanidin-3-glucoside by mass spectroscopy. The presence of cyanidin 3-rutinoside was confirmed by ^{1}H-NMR and ^{13}C-NMR.

Hydroxycinnamic acid and derivatives, such as chlorogenic acid are traditionally isolated by liquid-liquid extractions and crystallizations [71, 72]. However, these procedures are long, laborious and consume large quantities of organic solvents. For the isolation of these compounds at the laboratory level, chromatographic methods have some advantages, such as the speed and ease of scaling and automation. In a preliminary study, Rodríguez-Durán et al. [73] isolated by chlorogenic acid from CP by column chromatography. Chlorogenic acid was extracted with aqueous ethanol (50% v/v) by a microwave-assisted extraction technique. The crude extract was fractionated first, on an Amberlite XAD-16 column and then on a semi-preparative C18 HPLC column. This procedure led to a chlorogenic acid rich fraction with a purity of 61.7% purity and a recovery yield of 43.4%.

Several authors have purified chlorogenic acid from other natural sources using chromatographic techniques. For example, Zhang, et al. [74] developed a preparative separation method for the recovery of chlorogenic acid from crude extracts of *Lonicera japonica* by its adsorption on HPD–850 macroporous resin. Under optimum conditions, the chlorogenic acid content of the extract was increased from 11.2 to 50% with a recovery of 87.9%. Lu et al. [75] purified chlorogenic acid from a *Lonicera japonica* extract using high-speed counter-current chromatography (HSCCC) with a biphasic system composed of *n*-butanol-acetic acid-water (4:1:5). They obtained 16.9 mg of chlorogenic acid with a purity of 94.8% from 300 mg of crude extract, with a recovery of approximately 90%. HSCCC chromatography was also used to simultaneously purify caffeic acid, chlorogenic acid, and luteolin from a crude extract of *Caulis lonicerae*. They recovered 7.2 mg of caffeic acid, 15.7 mg of chlorogenic acid and 18.8 mg of luteolin from 110 mg of crude extract with a purity of 95.55%, 97.24%, and 98.11%, respectively [76].

5.3.3 ENZYMATIC TRANSFORMATION

Flavonoids and phenolic acids present in CP are natural compounds with important biological activities. However, because of their polar character, the applications of these polyphenols in dispersed lipid systems are limited. Chemical or enzymatic lipophilization of the polyphenolic skeleton may not only increase their solubility and stability in a lipophilic environment but also their biological properties. Lipophilization consists of the esterification of a hydroxyl or carboxyl group of the polyphenol with a lipophilic substituent [77, 78]. Enzymatic lipophilization has some advantages over chemical lipophilization, such as milder reaction conditions, minimization of side reactions and formation of byproducts, a selective specificity, a wider variety of pure synthetic substrates, fewer intermediary and purification steps, lower energy consumption and waste materials production [77].

Some authors have studied the lipophilization of phenolic compounds such as those present in CP. For example, Guyot et al. [79] studied the enzymatic synthesis of esters of phenolic acids with fatty alcohols catalyzed by an immobilized lipase of *Candida antarctica* in a solvent-free system. They observed very low reaction rates despite the high activity of the catalyst due to the inhibition of *C. antarctica* lipase by the presence of a conjugated double bond in the side chain of a *p*-hydroxylated aromatic ring in caffeic acid and ferulic acid.

Hernández et al. studied esterification of chlorogenic acid and 1-heptanol by immobilized *C. antarctica* lipase B in a supercritical CO_2/t-butanol system, while simultaneously extracting the formed compound. Under the optimal conditions (150 bar, 55°C, 10% t-butanol (v/v), 20 mg/mL of lipase and 20 mg/mL of molecular sieves (3Å)) the chlorogenic acid from a CP aqueous-methanolic extract, was selectively esterified with a conversion to heptyl ester of 65% in 25 h of reaction.

An alternative for the enzymatic synthesis of hydroxycinnamic acid esters is the use of a feruloyl esterase as a catalyst. Giuliani, et al. [80] studied the synthesis of pentyl ferulate catalyzed by a feruloyl esterase from *Aspergillus niger* in a water-in-oil microemulsion system. The reaction achieved a high yield (50–60%) in a short time (8 h). Feruloyl esterases have also been used in the synthesis of esters of hydroxycinnamic acids with monosaccharides [81, 82], oligosaccharides [83] and glycerol [84, 85].

Another potential application for CP polyphenols is the synthesis of caffeic acid phenethyl ester (CAPE) from chlorogenic or caffeic acid present in this by-product. CAPE is a potent anticancer component of propolis and possesses important anti-inflammatory, neuroprotective, hepatoprotective and cardioprotective activities [86].

Currently, CAPE is extracted from natural sources, such as propolis. The process is simple but slow and costly. Chemical synthesis by esterification between caffeic acid and phenethyl alcohol is slow (up to 4 days) and with low yields (40–60%) [87]. Enzymatic synthesis catalyzed by immobilized lipase of *C. antarctica* has high yields (90–100%). However, the reaction is slow (48–60 h) and requires large amounts of a biocatalyst (15–18 mg/mg caffeic acid) [88].

On the other hand, the transesterification reaction between chlorogenic acid and phenethyl alcohol catalyzed by an *Aspergillus japonicus* chlorogenate hydrolase leads to moderate yields (50% in molar basis) in a short reaction time (4 h) [89]. Olguín-Gutiérrez [90] studied the synthesis of CAPE by transesterification between chlorogenic acid and phenethyl alcohol catalyzed by an enzymatic extract of *Aspergillus ochraceus* in a diisopropyl ether-water biphasic system. In this system, a conversion of 9.4% was obtained in 12 h of reaction. At the same reaction conditions but using a CP extract as a substrate, no CAPE formation was detected until 168 h of reaction, probably due to the impurities present in the coffee extract.

5.4 CONCLUSIONS

Coffee pulp is a highly underutilized by-product. Using conventional separation techniques, valuable phenolic compounds can be obtained from CP, such as anthocyanins, proanthocyanidins, hydroxycinnamic acids and their derivatives. Enzymatic hydrolysis and solid-medium fermentation may enhance or facilitate the extraction of these compounds.

In addition, by means of enzymatic reactions in non-aqueous media, it is possible to modify the CP polyphenols to obtain phenolic compounds with improved solubility and biological activities, such as lipophilized phenolic acids and flavonoids, as well as potent bioactive compounds such as the phenethyl ester of caffeic acid. The exploitation of CP polyphenols could generate an added value to this byproduct and can help reduce the pollution generated by the coffee industry.

KEYWORDS

- antioxidants
- coffee pulp
- phenolic compounds

REFERENCES

1. International Coffee Organization: Total production by all exporting countries. http://www.ico.org/historical/1990%20onwards/PDF/1a-total-production.pdf (accessed July 12, 2017).
2. Valdez-Vazquez, I., Acevedo-Benitez, J. A., & Hernandez-Santiago, C., (2010). Distribution and potential of bioenergy resources from agricultural activities in Mexico. *Renew. Sust. Energ. Rev.*, *14*, 2147–2153.

3. Esquivel, P., & Jiménez, V. M., (2012). Functional properties of coffee and coffee by-products. *Food Res. Int.*, *46*, 488–495.
4. Aranda, E., Duran, L., & Escamilla, E., (2004). Vermicomposting in coffee cultivation. In: Wintgens, J. N., (ed.), *Coffee: Growing, Processing, Sustainable Production* (pp. 324–338). Wiley: Weinheim, Germany.
5. Noriega-Salazar, A., Silva-Acuña, R., & García de Salcedo, M., (2008). Use of coffee pulp in animal feed. *Zootec. Trop.*, *26*, 411–419.
6. Martínez-Carrera, D., Aguilar, A., Martínez, W., Bonilla, M., Morales, P., & Sobal, M., (2000). Commercial production and marketing of edible mushrooms cultivated on coffee pulp in Mexico. In: Sera, T., Soccol, C. R., Pandey, A., & Roussos, S., (eds.), *Coffee Biotechnology and Quality* (pp. 471–488). Kluwer Academic Publishers: Dordrecht, The Netherlands.
7. Pierre, F., Rosell, M., Quiroz, A., & Granda, Y., (2009). Chemical and biological evaluation of coffee pulp compost in caspito municipality Andrés Eloy Blanco, Estado Lara, Venezuela. *Bioagro, 21*, 105–110.
8. Agrifood and Fisheries Information Service: Statistical Yearbook of Agricultural Production. http://infosiap.siap.gob.mx/aagricola_siap_gb/ientidad/index.jsp (accessed July 12, 2017).
9. Ulloa, J. B., Van Weerd, J. H., Huisman, E. A., & Verreth, J. A. J., (2004). Tropical agricultural residues and their potential uses in fish feeds: The Costa Rican situation. *Waste Manage*, *24*, 87–97.
10. Bressani, R., (1979). The by-products of coffee berries. In: Braham, J. E., & Bressani, R., (eds.), *Coffee Pulp: Composition, Technology, and Utilization* (pp. 5–10). International Development Research Centre: Ottawa, Canadá.
11. Agrifood and Fisheries Information Service (2005). *Master plan for the coffee product system in Mexico* (p. 90). Ministry of Agriculture, Livestock and Animal Resources: México, D. F.
12. Aranda, E., & Barois, I., (2000). Coffee pulp vermicomposting treatment. In: Sera, T., Soccol, C. R., Pandey, A., & Roussos, S., (eds.), *Coffee Biotechnology and Quality* (pp. 489–506), Springer Netherlands: Dordrecht.
13. Ramírez-Martínez, J. R., (1999). *Silage Coffee Pulp.* National Experimental University of Táchira:: San Cristóbal, Venezuela, p. 156.
14. Musisi, A., (2009). Fuel from waste fires up Uganda. *Appropriate Technol., 36*, 33–35.
15. Shenoy, D., Pai, A., Vikas, R. K., Neeraja, H. S., Deeksha, J. S., Nayak, C., et al., (2011). A study on bioethanol production from cashew apple pulp and coffee pulp waste. *Biomass Bioenerg., 35*, 4107–4111.
16. Calzada, J. F., León, O. R., Arriola, M. C., Micheo, F., Rolz, C., León, R., & Menchú, J. F., (1981). Biogas from coffee pulp. *Biotechnol. Lett., 3*, 713–716.
17. Medeiros, A. B. P., Christen, P., Roussos, S., Gern, J. C., & Soccol, C. R., (2003). Coffee residues as substrates for aroma production by *Ceratocystis fimbriata* in solid state fermentation. *Braz. J. Microbiol., 34*, 245–248.
18. Torres-Mancera, M. T., Cordova-López, J., Rodríguez-Serrano, G., Roussos, S., Ramírez-Coronel, M. A., Favela-Torres, E., et al., (2011). Enzymatic extraction of hydroxycinnamic acids from coffee pulp. *Food Technol. Biotechnol., 49*, 369–373.

19. Murthy, P. S., Naidu, M. M., & Srinivas, P., (2009). Production of alpha-amylase under solid-state fermentation utilizing coffee waste. *J. Chem. Technol. Biotechnol., 84,* 1246–1249.
20. Murthy, P. S., & Naidu, M. M., (2011). Production and application of xylanase from *Penicillium* sp. Utilizing coffee by–products. *Food Bioprocess Technol., 5,* 657–664.
21. Boccas, F., Roussos, S., Gutierrez, M., Serrano, L., & Viniegra, G. G., (1994). Production of pectinase from coffee pulp in solid state fermentation system: Selection of wild fungal isolate of high potency by a simple three–step screening technique. *J. Food Sci. Technol., 31,* 22–26.
22. Mata, G., Hernandez, D. M. M., & Andreu, L. G. I., (2005). Changes in lignocellulolytic enzyme activites in six *Pleurotus* spp. strains cultivated on coffee pulp in confrontation with *Trichoderma* spp. *World J. Microbiol. Biotechnol., 21,* 143–150.
23. Pérez-Morales, G. G., Ramírez-Coronel, A., Guzmán-López, O., Cruz-Sosa, F., Perraud-Gaime, I., Roussos, S., et al., (2011). Feruloyl esterase activity from coffee pulp in solid-state fermentation. *Food Technol. Biotechnol., 49,* 352–358.
24. Adachi, O., Ano, Y., Akakabe, Y., Shinagawa, E., & Matsushita, K., (2008). Coffee pulp koji of *Aspergillus sojae* as stable immobilized catalyst of chlorogenate hydrolase. *Appl. Microbiol. Biotechnol., 81,* 143–151.
25. Sobal, M., Martinez-Carrera, D., Rio, B., & Roussos, S., (2003). Screening of edible mushrooms for polyphenol degradation and tannase production from coffee pulp and coffee husk. In: Roussos, S., Soccol, C. R., Pandey, A., & Augur, C., (eds.), *New Horizons in Biotechnology* (pp. 89–95), Kluwer Academic Press: Dordrecht, The Netherlands.
26. Perraud-Gaime, I., & Roussos, S., (1997). Selection of filamentous fungi for decaffeination of coffee pulp in solid state fermentation prior to formation of conidiospores. In: Roussos, S., Lonsane, B. K., Raimbault, M., & Viniegra-Gonzalez, G., (eds.), *Advances in Solid State Fermentation,* Springer Dordrecht, the Netherlands, pp. 209–221.
27. Bressani, R., (1979). Potential uses of coffee-berry by-products. In: Braham, J. E., & Bressani, R., (eds.), *Coffee Pulp: Composition, Technology, and Utilization* (pp. 17–24), International Development Research Centre: Ottawa, Canadá.
28. Houbron, E., Cano-Hernández, V., Reyes-Alvarado, L. C., & Rustrian, E., (2007). In search of a sustainable solution for the treatment of coffee waste. *Gaceta de la Universidad Veracruzana, 101.*
29. Pandey, A., Soccol, C. R., Nigam, P., Brand, D., Mohan, R., & Roussos, S., (2000). Biotechnological potential of coffee pulp and coffee husk for bioprocesses. *Biochem. Eng. J., 6,* 153–162.
30. Choi, I. S., Wi, S. G., Kim, S. B., & Bae, H. J., (2012). Conversion of coffee residue waste into bioethanol with using popping pretreatment. *Bioresour. Technol., 125,* 132–137.
31. Menezes, E. G. T., Do Carmo, J. R., Alves, J. G. L. F., Menezes, A. G. T., Guimarães, I. C., Queiroz, F., et al., (2014). Optimization of alkaline pretreatment of coffee pulp for production of bioethanol. *Biotechnol. Prog., 30,* 451–462.
32. Martinez-Carrera, D., Aguilar, A., Martinez, W., Morales, P., Sobal, M., Bonilla, M., & Larque-Saavedra, A., (1998). A sustainable model for rural production of edible mushrooms in Mexico. *Micol. Neotrop. Apl., 11,* 77–96.

33. Niladevi, K. N., Sukumaran, R. K., & Prema, P., (2007). Utilization of rice straw for laccase production by *Streptomyces psammoticus* in solid-state fermentation. *J. Ind. Microbiol. Biotechnol.*, *34*, 665–674.
34. Nava, I., Favela-Torres, E., & Saucedo-Castañeda, G., (2011). Effect of mixing on the solid-state fermentation of coffee pulp with *Aspergillus tamarii*. *Food Technol. Biotechnol.*, *49*, 391–385.
35. Ramirez-Coronel, M. A., Marnet, N., Kolli, V. S. K., Roussos, S., Guyot, S., & Augur, C., (2004). Characterization and estimation of proanthocyanidins and other phenolics in coffee pulp (*Coffea arabica*) by thiolysis-high-performance liquid chromatography. *J. Agric. Food Chem.*, *52*, 1344–1349.
36. Torres-Mancera, M. T., (2013). Enzymatic extraction of chlorogenic acid from coffee pulp. PhD Dissertation, Metropolitan Autonomous University, México City, Mexico.
37. Arellano-González, M. A., Ramírez-Coronel, M. A., Torres-Mancera, M. T., Pérez-Morales, G. G., & Saucedo-Castañeda, G., (2011). Antioxidant activity of fermented and nonfermented coffee (*Coffea arabica*) pulp extracts. *Food Technol. Biotechnol.*, *49*, 374–378.
38. Elias, L. G., (1979). Chemical composition of coffee-berry by-products. In: Braham, J. E., & Bressani, R., (Eds.), *Coffee Pulp: Composition, Technology, and Utilization* (pp. 11–16). International Development Research Centre: Ottawa, Canada.
39. Awolumate, E., (1983). Chemical composition and potential uses of processing wastes from some Nigerian cash crops. *Turrialba*, *33*, 381.
40. Peñaloza, W., Molina, M. R., Brenes, R. G., & Bressani, R., (1985). Solid-state fermentation: An alternative to improve the nutritive value of coffee pulp. *Appl. Environ. Microbiol.*, *49*, 388–393.
41. Donkoh, A., Atuahene, C. C., Kese, A. G., & Mensah-Asante, B., (1988). The nutritional value of dried coffee pulp (DCP) in broiler chickens' diets. *Anim. Feed Sci. Technol.*, *22*, 139–146.
42. Zuluaga-Vasco, J., (1990). In: *Proceedings of the I Seminario Internacional on Biotechnology in the Coffee Agribusiness Industry*, Xalapa, Mexico, Mexican Coffee Institute and Universidad Autonoma Metropolitana: Mexico City, Mexico.
43. Ulloa Rojas, J. B., Verreth, J. A. J., Van Weerd, J. H. Huisman, E. A., (2002). Effect of different chemical treatments on nutritional and antinutritional properties of coffee pulp. *Anim. Feed Sci. Technol.*, *99*, 195–204.
44. Ramirez-Martinez, J. R., (1988). Phenolic compounds in coffee pulp: Quantitative determination by HPLC. *J. Sci. Food Agric.*, *43*, 135–144.
45. Clifford, M. N., & Ramirez-Martinez, J. R., (1991). Phenols and caffeine in wet-processed coffee beans and coffee pulp. *Food Chem.*, *40*, 35–42.
46. Ramírez-Coronel, M. A., Torres-Mancera, M. T., Augur, C., & Saucedo-Castañeda, G., (2007). In: *Book of Abstracts*, XII National Congress of Biotechnology and Bioengineering, Morelia, México, Mexican Society of Biotechnology and Bioengineering: Morelia, Mexico.
47. Benoit, I., Navarro, D., Marnet, N., Rakotomanomana, N., Lesage-Meessen, L., Sigoillot, J. C., et al., (2006). Feruloyl esterases as a tool for the release of phenolic compounds from agro-industrial by-products. *Carbohydr. Res.*, *341*, 1820–1827.

48. Clifford, M. N., & Ramirez-Martinez, J. R., (1991). Tannins in wet-processed coffee beans and coffee pulp. *Food Chem.*, *40*, 191–200.
49. De Colmenares, N. G., Ramírez-Martínez, J. R., Aldana, J. O., & Clifford, M. N., (1994). Analysis of proanthocyanidins in coffee pulp. *J. Sci. Food Agric.*, *65*, 157–162.
50. Rodríguez-Durán, L. V., Ramírez-Coronel, M. A., Aranda-Delgado, E., Nampoothiri, K. M., Favela-Torres, E., Aguilar, C. N., et al., (2014). Soluble and bound hydroxycinnamates in coffee pulp (*Coffea arabica*) from seven cultivars at three ripening stages. *J. Agric. Food Chem.*, *62*, 7869–7876.
51. Kumar, S., & Pandey, A. K., (2013). Chemistry and biological activities of flavonoids: An overview. *Sci. World, J.*, 1–16.
52. Crozier, A., Jaganath, I. B., & Clifford, M. N., (2006). Phenols, polyphenols and tannins: An overview. In: Crozier, A., Clifford, M. N., & Ashihara, H., (eds.), *Plant Secondary Metabolites* (pp. 1–24). Blackwell Publishing Ltd: Oxford, U. K.
53. Ramírez, J., (1987). Phenolic compounds in coffee pulp. Fresh pulp paper chromatography of 12 cultivars of *Coffea arabica* L. *Turrialba*, *37*, 317–323.
54. Robbins, R. J., (2003). Phenolic acids in foods: An overview of analytical methodology. *J. Agric. Food Chem.*, *51*, 2866–2887.
55. Marques, V., & Farah, A., (2009). Chlorogenic acids and related compounds in medicinal plants and infusions. *Food Chem.*, *113*, 1370–1376.
56. Chen, Y., Jimmy, Yu, Q., Li, X., Luo, Y., & Liu, H., (2007). Extraction and HPLC characterization of chlorogenic acid from tobacco residuals. *Sep. Sci. Technol.*, *42*, 3481–3492.
57. Augur, C., Gutierrez-Sanchez, G., Ramirez-Coronel, A., Contreras-Dominguez, M., Perraud-Gaime, I., & Roussos, S., (2006). Analysis of antiphysiological components of coffee pulp. In: Laroche, C., Pandey, A., & Dussap, C. G., (eds.), *Current Topics in Bioprocesses in Food Industries*, (Vol. I, pp. 391–402). Asiatech: New Delhi, India.
58. Koleckar, V., Kubikova, K., Rehakova, Z., Kuca, K., Jun, D., Jahodar, L., et al., (2008). Condensed and hydrolyzable tannins as antioxidants influencing the health. *Mini Rev. Med. Chem.*, *8*, 436–447.
59. Garcia, L., Velez, A., & De Rozo, M., (1985). Extraction and quantification of polyphenols from coffee pulp. *Arch. Latinoam. Nutr.*, *35*, 491–495.
60. Tobón, A. N. D. L. C., (2015). Ultrasound-assisted extraction of phenolic compounds from coffee pulp (*Coffea arabica* L.) variety Castillo. MSc Dissertation, Lasallian University Corporation, Caldas, Colombia.
61. Rodríguez-Durán, L. V., Ortega-Hernández, A. K., Ramírez-Coronel, M. A., Favela-Torres, E., Aguilar-González, C. N., & Saucedo-Castañeda, G., (2014). In: *Book of Abstracts, 6th Food Science, Biotechnology and Safety Meeting*, Monterrey, Mexico, Mexican Association of Food Science A.C.: Monterrey, Mexico.
62. Nardini, M., Cirillo, E., Natella, F., Mencarelli, D., Comisso, A., & Scaccini, C., (2002). Detection of bound phenolic acids: Prevention by ascorbic acid and ethylenediaminetetra acetic acid of degradation of phenolic acids during alkaline hydrolysis. *Food Chem.*, *79*, 119–124.
63. Krygier, K., Sosulski, F., & Hoggs, L., (1982). Free, esterified, and insoluble-bound phenolic acids. 1. Extraction and purification procedure. *J. Agric. Food Chem.*, *30*, 330–334.

64. Nieter, A., Kelle, S., Linke, D., & Berger, R. G., (2016). Feruloyl esterases from *Schizophyllum commune* to treat food industry side-streams. *Bioresour. Technol.*, *220*, 38–46.
65. Torres-Mancera, M. T., Baqueiro-Peña, I., Figueroa-Montero, A., Rodríguez-Serrano, G., González-Zamora, E., Favela-Torres, E., et al., (2013). Biotransformation and improved enzymatic extraction of chlorogenic acid from coffee pulp by filamentous fungi. *Biotechnol. Prog.*, *29*, 337–345.
66. Ramirez-Velasco, L., Armendariz-Ruiz, M. A., Arrizon, J., Rodriguez-Gonzalez, J. A., & Mateos-Diaz, J. C., (2016). Liberation of caffeic acid from coffee pulp using an extract with chlorgenate esterase activity of *Aspergillus ochraceus* produced by solid state fermentation. *Rev. Mex. Ing. Quim.*, *15*, 503–512.
67. Bhoite, R. N., Navya, P. N., & Murthy, P. S., (2013). Statistical optimization of bioprocess parameters for enhanced Gallic acid production from coffee pulp tannins by *Penicillium verrucosum*. *Prep. Biochem. Biotechnol.*, *43*, 350–363.
68. Palomino, G. L. R., Biasetto, C. R., Araujo, A. R., & Bianchi, V. L. D., (2015). Enhanced extraction of phenolic compounds from coffee industry s residues through solid state fermentation by *Penicillium purpurogenum*. *Food Sci. Technol. (Campinas)*, *35*, 704–711.
69. De Colmenares, N. G., Ramírez-Martínez, J. R., Aldana, J. O., Ramos-Niño, M. E., Clifford, M. N., Pékerar, S., et al., (1998). Isolation, characterisation and determination of biological activity of coffee proanthocyanidins. *J. Sci. Food Agric.*, *77*, 368–372.
70. Murthy, P. S., Manjunatha, M. R., Sulochannama, G., & Naidu, M. M., (2012). Extraction, characterization and bioactivity of coffee anthocyanins. *Eur. J. Biol. Sci.*, *4*, 13–19.
71. Hulme, A. C., (1953). The isolation of chlorogenic acid from the apple fruit. *Biochem J.*, *53*, 337–340.
72. Rakotomalala, J. J., (1992). Biochemical diversity of coffee trees: Analysis of hydroxycinnamic acids, purine bases and glycosidic diterpenes, particularity of wild coffee trees in the Malagasy region (*Mascarocoffea* Chev.). PhD thesis, University of Montpellier II, Montpellier, France.
73. Rodríguez-Durán, L. V., Ramírez-Coronel, M. A., Favela-Torres, E., Aguilar-González, C., & Saucedo-Castañeda, G., (2016). In: *Book of Abstracts, 7th Food Science, Biotechnology and Safety Meeting*, Cancun, Mexico, Asociación Mexicana de Ciencia de los Alimentos A.C.: Cancun, Mexico.
74. Zhang, B., Yang, R., Zhao, Y., & Liu, C. Z., (2008). Separation of chlorogenic acid from honeysuckle crude extracts by macroporous resins. *J. Chromatogr. B*, *867*, 253–258.
75. Lu, H. T., Jiang, Y., & Chen, F., (2004). Application of preparative high-speed counter-current chromatography for separation of chlorogenic acid from *Flos Lonicerae*. *J. Chromatogr. A*, *1026*, 185–190.
76. Wang, Z., Wang, J., Sun, Y., Li, S., & Wang, H., (2008). Purification of caffeic acid, chlorogenic acid and luteolin from *Caulis Lonicerae* by high-speed counter-current chromatography. *Sep. Purif. Technol.*, *63*, 721–724.

77. Figueroa-Espinoza, M. C., & Villeneuve, P., (2005). Phenolic acids enzymatic lipophilization. *J. Agric. Food Chem.*, *53*, 2779–2787.
78. Danihelová, M., Viskupičová, J., & Šturdík, E., (2012). Lipophilization of flavonoids for their food, therapeutic and cosmetic applications. *Acta Chim. Slov.*, *5*, 59–69.
79. Guyot, B., Bosquette, B., Pina, M., & Graille, J., (1997). Esterification of phenolic acids from green coffee with an immobilized lipase from *Candida antarctica* in solvent-free medium. *Biotechnol. Lett.*, *19*, 529–532.
80. Giuliani, S., Piana, C., Setti, L., Hochkoeppler, A., Pifferi, P. G., Williamson, G., et al., (2001). Synthesis of pentylferulate by a feruloyl esterase from *Aspergillus niger* using water-in-oil microemulsions. *Biotechnol. Lett.*, *23*, 325–330.
81. Topakas, E., Vafiadi, C., Stamatis, H., & Christakopoulos, P., (2005). *Sporotrichum thermophile* type C feruloyl esterase (StFaeC): Purification, characterization, and its use for phenolic acid (sugar) ester synthesis. *Enzyme Microb. Technol.*, *36*, 729–736.
82. Couto, J., Karboune, S., & Mathew, R., (2010). Regioselective synthesis of feruloylated glycosides using the feruloyl esterases expressed in selected commercial multi-enzymatic preparations as biocatalysts. *Biocatal. Biotransform.*, *28*, 235–244.
83. Couto, J., St-Louis, R., & Karboune, S., (2011). Optimization of feruloyl esterase-catalyzed synthesis of feruloylated oligosaccharides by response surface methodology. *J. Mol. Catal. B: Enzym.*, *73*, 53–62.
84. Tsuchiyama, M., Sakamoto, T., Tanimori, S., Murata, S., & Kawasaki, H., (2007). Enzymatic synthesis of hydroxycinnamic acid glycerol esters using type A feruloyl esterase from *Aspergillus niger*. *Biosci. Biotechnol. Biochem.*, *71*, 2606–2609.
85. Matsuo, T., Kobayashi, T., Kimura, Y., Tsuchiyama, M., Oh, T., Sakamoto, T., et al., (2008). Synthesis of glyceryl ferulate by immobilized ferulic acid esterase. *Biotechnol. Lett.*, *30*, 2151–2156.
86. Tolba, M. F., Azab, S. S., Khalifa, A. E., Abdel-Rahman, S. Z., & Abdel-Naim, A. B., (2013). Caffeic acid phenethyl ester, a promising component of propolis with a plethora of biological activities: A review on its anti-inflammatory, neuroprotective, hepatoprotective, and cardioprotective effects. *IUBMB Life*, *65*, 699–709.
87. Zhang, P., Tang, Y., Li, N. G., Zhu, Y., & Duan, J. A., (2014). Bioactivity and chemical synthesis of caffeic acid phenethyl ester and its derivatives. *Molecules*, *19*, 16458.
88. Widjaja, A., Yeh, T. H., & Ju, Y. H., (2008). Enzymatic synthesis of caffeic acid phenethyl ester. *J. Chin. Inst. Chem. Eng.*, *39*, 413–418.
89. Kishimoto, N., Kakino, Y., Iwai, K., & Fujita, T., (2005). Chlorogenate hydrolase-catalyzed synthesis of hydroxycinnamic acid ester derivatives by transesterification, substitution of bromine, and condensation reactions. *Appl. Microbiol. Biotechnol.*, *68*, 198–202.
90. Olguín-Gutiérrez, J. S., (2010). Synthesis of esters of brownic acid by enzymatic route. MSc Dissertation, University of Guadalajara, Guadalajara, Mexico.

CHAPTER 6

ESSENTIAL OILS IN ACTIVE FOOD PACKAGING

OLGA B. ALVAREZ-PÉREZ,[1] MÓNICA L. CHÁVEZ-GONZÁLEZ[2],
ANNA ILINÁ,[2] JOSÉ LUIS MARTÍNEZ-HERNÁNDEZ,[2]
ELDA PATRICIA SEGURA-CENICEROS,[2]
RODOLFO RAMOS-GONZÁLEZ[3] and CRISTÓBAL N. AGUILAR[1]

[1] *Autonomous University of Coahuila, Food Research Department, School of Chemistry, Blvd. V. Carranza corner José Cárdenas Valdés s/n Col, República Oriente, ZIP 25280, Saltillo, Coahuila, México*

[2] *Autonomous University of Coahuila, Nanobioscience Group, School of Chemistry, Blvd. V. Carranza corner José Cárdenas Valdés s/n Col, República Oriente, ZIP 25280, Saltillo, Coahuila, México, E-mail: monicachavez@uadec.edu.mx*

[3] *CONACYT – Autonomous University of Coahuila, Blvd. V. Carranza corner José Cárdenas Valdés s/n Col, República Oriente, ZIP 25280, Saltillo, Coahuila, MéxicoAbstract*

ABSTRACT

The current way of life, environmental care policies, search and use of biodegradable materials and the interest of consumers on natural food additives are reasons to make use of essential oils (EOs), they are an interesting ingredient to produce eco-friendly food packaging and they can confer functional properties to package like as antioxidant and antimicrobial, these properties can extend shelf life of foods and in this way increase the value of the products.

In this document shows the most recent reports on essential oils used in active food packaging, methods of adding oils, impact on several properties during formation of active package, antimicrobial and antioxidant action of EOs reported under different conditions.

6.1 INTRODUCTION

The current way of life and the demands of modern society have fostered the development of food technology, especially the area of food packaging. This area has improved a lot in the last years, trying to cover the needs of the contemporary consumer, where the search for fresh, slightly preserved and safe food, improve the quality and product with a longer shelf life are the main requirements in the food industry [1].

Nowadays, food packaging has evolved gradually from simple containers as a barrier between food and the external environment to preserve it to packaging with technologies to increase the safety and commercialization of the products through the use of eco-friendly materials. The active packaging is an innovative packaging systems that aims to improve food safety, preserve the quality of packaged food and extend its shelf life controlling the development of microorganisms avoid the decay and reduce the need to add preservatives directly to food [2, 3].

The incorporation of an active substance inside packaging can promote the effectiveness of food packaging and provide numerous functionalities for prolonging the shelf life of foods [4] and minimize food losses and wastage. That active character can be given by addition of essential oils converting antimicrobial packaging is an important type of active packaging. EO's are considered natural antimicrobials, also It has been determined that they are considered non-toxic, in fact nowadays, consumers prefer to purchase healthy and cleaner products which converts essential oils a powerful tool to preserve food [5, 6].

6.2 ESSENTIAL OILS

6.2.1 NATURE OF EOs

Essential oils (EOs) are liquid, natural, volatile and complex, they are a mixture of several chemicals biosynthesized by plants as secondary metabolites that are produced as a defense for plant against diverse pathogens, fungal, bacteria, viral, insects, herbivores among others [7]. They can be obtained from different parts of the plant-like flowers, leaves, seeds, barks, roots, woods, and fruits. Essential oils can be extracted through different such as distillation, hydrodistillation, stem distillation, cold pressing or maceration [8, 9].

The composition of EOs depends on species and type, part and mature stage of plant. However, the major component found in EOs are terpenes, in special the subclass monoterpenes and sesquiterpenes represent around 80–90% of total content of EOs. Also, there are present other compounds like esters, aldehydes, ketones, alcohols, acids, amines, sulfides, epoxides, and carbures [7, 10]. In general, the composition of EOs can be dived into two groups, terpenes hydrocarbons and aromatic compounds. In monoterpenes group, monocyclic and bicyclic can be found as cymene, α-pinene, β-pinene, sabinene; compounds of phenolic type as carvacrol and thymol, alcohols like as geraniol, citronellol. In the second group, aromatic compounds are constituted by aldehydes, alcohol like cinnamyl alcohol, phenol, metoxy derivatives as anethole, estragole, terpenoids as menthol and ascaridol [9, 11, 12]. Nevertheless, the exactly composition of any particular EOs depends no only of vegetable source but also of the part of plant from which the essential oil is extracted, also growth conditions of plant has an impact on composition of EOs [13]. EOs is commonly used as flavoring agents in different industries like medicine, perfumery, cosmetic, among others [14]. In addition to the peculiar odors of EOs, one of its most important characteristic is the high effective antimicrobial effect against large numerous of bacteria and fungi and variety of foodborne pathogens [15] also EOs have approbation by the Food and Drug Administration (FDA) for being used in food as GRAS (generally recognized as safe) [16]. EOs, a kind of natural substance with powerful antimicrobial activity against the wide variety of compounds present in essential oils makes them excellent antimicrobial agents with low risks of generating microbial resistance. The incorporation of EOs into food packaging material is an excellent option to prolong the shelf life of food due to the EOs can be released slowly as a vapor into the packaging headspace or by direct contact act as antimicrobial effect in the environment [16, 17].

6.2.2 MECHANISM OF ACTION

Antimicrobial action of EOs is due at the synergistic action of different compounds presents in it; in case of terpenes due to their non-polar character they are usually associated with lipids presents in microbial cells, EOs increases fluidity and permeability of microbial membrane caused disturbances in respiratory processes and eventually cause cell death [18].

Phenolic fraction of EOs acts on proteins present in cell membrane; they are binding to protein and causes disruption and loss of cellular function [19]. EO's are considered natural antimicrobials, efficient and current society prefers them to be greener.

6.2.3 SOURCES OF EOs

There are different sources of obtaining EOs, among the most used are citrus, oregano, pepper, cinnamon, rosemary, clove, lavender, garlic, basil, thyme and mint (Table 6.1). Yield in EOs extraction vary depends on vegetable source, harvesting time and although all parts of the plant contain their composition vary with the part of plant used as raw material [8]. Other factors such as cultivation, soil and climatic conditions and harvesting time can also determine the composition and quality of the essential oil, also the extraction method applied to recovery EOs has a strong impact to maintain the proportion of components of EOs [9].

TABLE 6.1   Main Constituents of Select Essential Oils Derived From Plants

Plant Source	Main components	Latin name	Reference
Rosemary	1,8-cineole Camphor α-Pinene	*Rosmarinus offcinalis* L.	[12]
Cinnamon	Trans-cinnemaldehyde Limonene Eugenol	*Cinnamomum verum*	[20]
Oregano	thymol α-Terpinene Carvacrol 4-Terpineol	*Origanum vulgare*	[21]
Pepper	D-germacrene sabinene a-pinene	*Schinus terebinthifolius* R.	[22]
Clove	Eugenol Caryophyllene α-caryophyllene 2-methoxy–4-(2-pro- penyl) phenol acetate caryophyllene oxide	*Syzygium aromaticum*	[23]

Essential Oils in Active Food Packaging

TABLE 6.1 *(Continued)*

Plant Source	Main components	Latin name	Reference
Ziziphora	Carvacrol Thymol p-Cymene γ-Terpinene	*Ziziphora clinopodioides*	[24]
Lavender	Linalool Camphor Terpinen–4-ol 1,8- cineol	*Lavandula officinalis*	[25]
Garlic	Diallyl trisulfide Methyl allyl trisulfide Diallyl disulfide	*Allium sativum*	[26]
Citrus	Limonene γ-terpinene β-pinene α-terpineol		[27]
Thyme	Thymol p-Cymene γ-Terpinene Thymyl acetate	*Thymus linearis* B	[28]
Mint	L-Menthone L-Menthol Menthofuran Caryophyllene	*Mentha* × *piperita* L	[29]
Basil	Chavicol methyl ether Linalool Cineole α-Bergamotene	*Ocimum basilicum*	[29]

6.3 INCORPORATION OF EOs IN PACKAGING

The incorporation of the antimicrobial agents into the packaged material can be in three different forms, (1) added directly to the food, (2) added in a separate container, and (3) incorporated into the packaging material [6]. When EOs are added directly, non-polar character of EOs show more affinity between EOs and components in food, mainly lipids [2], causing migration of bioactive components to food and conferring beneficial properties. When the Incorporation into packaging material: The addition of EOs into the packaging can be carried out adding them directly in mixture of polymers or through coating them onto surface [17]. The combination of EOs and packaging methods has been assessed in order to increase the shelf life.

EOs has been used in combination with different types of packing to protect a wide variety of foods; cinnamon EO has been tested on pork using modified atmosphere (MAP) and vacuum packaging and increased the shelf life of product causing antimicrobial effect against yeast and molds [30]; cinnamon has been applied in combination with oregano on maize in with ethylene-vinyl alcohol (EVOH) copolymer films and tested against *A. flavus, A. parasiticus* [31]; Wen et al. [15] reported the use of Cinnamon combined with nanofibrous electrospun polyvinyl alcohol to be added on strawberry, results showed an excellent action against *Staphylococcus aureus* and *Escherichia coli*. Tomato puree has been tested in a multi-layer material of polyurethane adhesive packaging added with cinnamon and also show an excellent antimicrobial activity against *Escherichia coli* O157:H7 and *Saccharomyces cerevisiae* [32] fresh beef also was tested in films of cast polypropylene/polyvinyl alcohol with cinnamon [33].

Oregano was used to make polyvinyl alcohol (PVA) film to cover cherry tomatoes [34] they used 2% of oregano essential oil and results showed effectiveness at suppressing the growth of microorganisms by diffusion of EOs. Paparella [35] reported the use of oregano in modified atmosphere on pork fillets and was effective against *L. monocytogenes, Pseudomonas* and *Brochothrix thermosphacta* even oregano combined with others antimicrobial compounds like as citral in an active bag of polypropylene and ethylene-vinyl alcohol copolymer (PP/EVOH) [36], caprylic acid in mix with oregano and vacuum packed also showed antimicrobial action against *Listeria monocytogenes* on minced beef [37]. Rosemary essential also has been tested against *Brochothrix, Pseudomonas* spp, *Enterobacteriaceae* in synergistic action with MAP [38]. Lavender essential oil was effective against *Botrytis cinerea*, it was tested in Strawberry EO was put it on sachet inside of clamshell [39]. Migration of EOs when are present in polymeric materials is by diffusion [40].

Active films of cellulose acetate in micro-atmosphere with pepper in its formulation was tested on Sliced mozzarella cheese and was effective against *Staphylococcus aureus* and *Listeria monocytogenes* [2] on the other hand, researchers tested action of clove in films of linear low-density polyethylene (LLDPE) modified with chromic acid they showed action against *Salmonella typhimurium* and *Listeria monocytogenes* on chicken [41] garlic at 2–8% in films of polyethylene tested in ready-to-eat beef loaves [42].

6.3.1 ESSENTIAL OILS INCORPORATED IN EDIBLE FILMS

Edible films are defined as continuous matrices that can be formed by proteins, polysaccharides, and lipids which allow the formulation of blends improving the quality or effectiveness. These systems have the possibility of incorporating essential oils and provide stability to the food or can be used as carriers of foods additives such as antioxidant or antimicrobial agents and a wide number of compounds, which can extend the product shelf life [43]. The incorporation of EOs in the films has advantage, mainly the reduction of undesirable interference in sensory characteristics; production of film allows a gradual release of active compounds that protects food during a longer time reducing the microbial contamination, and reduction of lipid oxidation and improve the safety and quality of food [6, 17]. The vast majority of films prepared are by casting method, that consists in evaporate the solvent from film formation solution until formatting a compact matrix [44].

Due to their lipidic nature in the last 30 years have been used in edible films with the aim to reduce water vapor permeability and also provide some properties (Table 6.2) [43].

6.4 EFFECTS OF EOs INCORPORATED IN EDIBLE FILMS

6.4.1 EFFECT ON PHYSICAL CHARACTERISTICS OF AN EDIBLE FILM INCORPORATED WITH OIL

Physical characteristics of edible films are of great importance to be able to know, or to predict their behavior when are applied to a food system. These properties could be affected by the incorporation of more components within its formulation, such as thickness, color, water activity, water vapor permeability, solubility among others. It is important the evaluation of optical, mechanical, and barrier properties of films like as tensile

TABLE 6.2 Essential Oils in Edible Active Packaging

Essential oil	Matrix	Oil:matrix proportion	Food System	Effectiveness	Reference
Clove	Sunflower protein	0.75 (v/v):1 (w/v)	Refrigerated sardine patties	Retard lipidic autooxidation and delay the growth of total mesophiles.	[45]
Almond and Walnut	Whey protein	0.5, 1.0 (v/v): 8 (w/v)		In emulsified films showed a decrease in water vapor permeability and increase in contact angle values.	[46]
Thai	Hidroxypropyl methylcellulose	1.5(w/v):2 (w/w)		Films showed antibacterial activity against *S. aureus* and *E. coli*, and also the mechanical and physical properties are improved.	[47]
Z. multiflora Boiss and *Mentha pulegium*	Corn zein	2,1(v/v):1.4 (w/v)		Antioxidant activity was improved, and films showed significant synergistic impact on microbial growth against *L. monocytogenes*, *S. aureus* and *E. coli*	[48]
Thyme, basil and oleic acid	Chitosan	1, 0.75. 0.5, 0.25 (v/v):1 (v/v)		Films did not inhibit the growth of *A. niger, B. cinerea* and *R.stolonifer* in an *in vitro* test, although show great properties as film.	[49]
Oregano and Bergamot	Hidroxypropyl methylcellulose	0.5, 1.0, 1.5. 2.0 (v/v): 4 (w/v)	*Prunus salicina L.*	Films were able to prolong shelf life for 14 days of storage compared to the control.	[50]
Corn, soybean, sunflower, and olive	Kappa carrageenan (*Euchema cotton*)	1–3 (v/v): 1(v/v)		Significantly reduced the moisture content and solubility in water and improved water vapor permeability.	[51]

TABLE 6.2 (Continued)

Essential oil	Matrix	Oil:matrix proportion	Food System	Effectiveness	Reference
Lemongrass (encapsulated)	Alginate	0.5 (w/v):1 (v/v)		Improved antimicrobial activity of active films, differed according to the type of microorganism. Film properties made them a great material for food packaging.	[52]
Myrcia ovata Cambessedes	Chitosan and cassava starch	0.5–2.5(w/v):0.4–1.6 (w/v):0.2–0.8 (v/v)	Mangaba	Films reduced the natural microflora (total aerobic mesophilic bacteria, molds and yeasts) and inhibited *B. cereus* (artificially contaminated) in mangabas stored at 10°C for 12 days.	[53]
Cinnamon, clove, and oregano	Starch-gelatin blend	0.25:1 (v/v)		Films showed antifungal activity against *Fusarium oxysporum, f.sp. gladiolo* and *Colletotrichum gloeosporoides* and also better functional properties compared with control.	[54]

strength, elongation at break, elastic properties, thermogravimetric, FT IR, and DSC analyses among others [51, 55, 56].

Hashemi et al. [57] investigated the addition of oregano essential oil in basil-seed gum films, they observed that thickness was recorded as 0.06 mm, and they do not found significantly difference when concentration of essential oil increased. In the moisture content was from 17.67 to 17.92% with an increment of oregano oil from 2–6%. Transparency was affected by the addition of essential oil, these films became more transparent, and for water vapor permeability when concentration increased the value of this parameter decreased from 0.426 to 0.369 × 10^{-10} g H_2O m^{-2} s MPa^{-1}.

Acosta et al. [54] studied starch-gelatin blends. They reported that structure is affected by the content of lipids in emulsion system. In barrier properties, addition of essential oils caused a decrease in water vapor permeability and also in mechanical properties (tensile strength, deformation at fracture point and elasticity modulus).

Acevedo-Fani, et al. [58] found that droplet size of nanoemulsions is affected by the addition of essential oil (Thyme, lemongrass and sage oil) into the alginate based edible films, suggesting different grades of affinity between the oil and alginate. Optical properties are also affected by the addition of essential oil, as well as water vapor permeability causing a decrease due to their hydrophobic behavior.

Klangmuang & Sothornvit [47] studying the properties of hydroxypropyl methylcellulose-based films with Thai essential oil, found that EO did not significantly affect the moisture content, but also this parameter decrease with EO addition, same behavior is observed in water vapor permeability. As expected, color is affected by EO.

Effect of essential oil addition on the physical properties of an edible film is variable and depends on the specific interactions between the oil components and the polymer matrix [43] is preferably that effectiveness of the film be tested on food [6].

6.4.2 EFFECT ON ANTIBACTERIAL ACTIVITY OF AN EDIBLE FILM INCORPORATED WITH ESSENTIAL OIL

Antibacterial activity is also important when essential oils are incorporated in edible films. Some studies could be observed in Table 6.3. Some phenolic compounds such as carvacrol, thymol, terpinene, and p-cymene,

Essential Oils in Active Food Packaging

TABLE 6.3 Antimicrobial Activity of Essential Oils in Edible Films

Essential oil	Concentration	Microorganism tested	Effectiveness	Reference
Cinnamon, lemongrass and oregano	1–10%	*Botrytis* sp., *Pilidiella granati* and *Penicillium* sp.	Films enriched with either oregano or cinnamon exhibited complete inhibition against *Botrytis* sp. *Penicillium* sp. and *P. granati* sp. at 10 g L^{-1}.	[59]
Oregano	1–6%	*S. Typhimurium E. coli* O157:H7, *P. aeruginosa, S. aureus B. cereus*	*B. cereus* and *S. aureus* were found to be more susceptible against BSG film containing essential oil	[57]
Lime	0.05–0.1%	*E. coli, S. typhymuium, B. cereus, S. aureus* and *L. monocytogenes*	For *E. coli, S. aureus* and *L. monocytogenes*, 500 ppm of essential oil added to bagasse edible film was enough to inhibit their growth.	[60]
Oregano and lavender	0.2%, 0.6%	*E. coli, S. aureus*	Greater inhibition of oils and pure compounds was observed against *S. aureus* (1000–2000 μg/mL MIC).	[61]
Thyme, lemongrass and sage	1%	*E. coli*	Thyme essential oil nanoemulsion showed an inhibitory effect significantly strong than lemon grass and sage, which do not any growth reduction.	[58]
Marjoram, coriander and clove	0.25%	*E. coli* *S. aureus* *S. enteritidis*	For *S. enteritidis*, the populations decreased to 4.72, 5.14, and 4.50 log CFU/mL after 12 h, *S. enteritidis* slightly increased for 3 and 9 h and for *L. monocytogenes* continuously decrease	[62]
Oregano, rosemary and garlic	1–4%	*Lactobacillus plantarum* (DSM 20174), *Salmonella enteritidis* (ATCC 13076), *E. coli* O157:H7 (ATCC 35218), *Listeria monocytogenes* (NCTC 2167) and *Staphyloccocus aureus* (ATCC 43300)	Greatest zone of inhibition was observed at 4% level against *S. aureus* (43.07 mm^2), *S. enteritidis* (40.59 mm^2), and *L. monocytogenes* (41.65 mm^2).	[63]

as active compounds of essential oil, can be correlated as the responsible agents for antibacterial activity of essential oils [14].

The antimicrobial action of EOs may hinder the development of pathogenic microorganisms and subsequent deterioration of food, microorganisms can accelerate the oxidation or/and produce changes in sensory properties.

6.4.3 EFFECT ON ANTIOXIDANT PROPERTY OF AN EDIBLE FILM WITH ESSENTIAL OILS

The content of phenolic compounds present in essential oils is directly correlated with antioxidant power [14]. Oxidation is an important way of food degradation and is produced during the storage of food products, oxidation is responsible for changes in the sensory attributes. Lipid oxidation also to produce rancid odors and flavors and a decrease in nutritional quality and the formation of potentially toxic compounds. The chemical composition of EOs makes their in excellent antioxidants [5]. Some studies demonstrated the antioxidant power of some edible film systems incorporated with different essential oils. Most of them mention that antioxidant power is due to the polyphenol content in essential oils [64] (Table 6.4).

6.5 CONCLUSIONS

Preservation of food is a necessity, so the development of packaging that helps maintain its integrity, quality, hygiene and prolong shelf life is an important task. The integration of compounds with bioactivities within food packaging, such as essential oils, is a promising task. Essential oils have shown excellent properties as natural antimicrobials that can be added to different matrices for the production of active food packaging, also use of EOs can reduce the use of synthetic additives. The action of EOs occurs by diffusion of compounds of EOs incorporated into food packaging, and its incorporation promotes the antimicrobial properties, the effectiveness of active packaging against microorganisms depends on nature, type, concentration of essential oil and microorganism tested.

Essential Oils in Active Food Packaging 143

TABLE 6.4 Antioxidant Activity of Essential Oils in Edible Films

Essential oil	System	Concentration	Antioxidant test	Effectiveness	Reference
Oregano	Basil-seed gum	1-6%	DPPH, ABTS and FRAP	DPPH – 0.14–0.71 g/kg ABTS – 4.56-36.84 g/kg FRAP – 0.46–3.14 g/kg	[57]
Eucalyptus globulus	Chitosan	0 – 4%	DPPH Nitric oxide radical scavenging assay and Hydrogen peroxide radical scavenging assay	DPPH scavenging activity of the films varied from 23.03% to 43.62%, for nitric oxide scavenging activity of the films varied from 35.23% to 70.47% and H_2O_2 varied from 27.4% to 63.15%	[64]
Oregano and lavender	Biogenic gelatin	0.2–0.6%	DPPH and FRAP	For DPPH results were from 12.7 – 60% foro regano oil and 1.5 to 8.6% for levander oil, similar resultas was observed for FRAP 94 – 241(AAE ppm/g film) and 0.1 – 26.0 (AAE ppm/g film) respectively.	[61]
Clove	Sunflower protein	7.5%	ABTS, FRAP and Photochemiluminescence (PCL)	ABTS 1194.1 (mg/g) FRAP 5733.3 (mmol/g) PCL 229 (µmol/g)	[45]
Citronella (*Pelargonium citrosum*), Coriander (*Coriandrum sativum*), Tarragon (*Artemisia dracunculus*) and thyme (*Thymus vulgaris*)	Hake protein	25%	DPPH	70–75%	[65]

KEYWORDS

- essential oils
- films
- packaging

REFERENCES

1. Ahmed, I., Lin, H., Zou, L., Brody, A. L., Li, Z., Qazi, I. M., et al., (2017). A comprehensive review on the application of active packaging technologies to muscle foods. *Food Control, 82*, 163–178.
2. Dannenberg, G. S., Funck, G. D., Cruxen, C. E. S., Marques, J. L., Silva, W. P., & Fiorentini, A. M., (2017). Essential oil from pink pepper as an antimicrobial component in cellulose acetate film: Potential for application as active packaging for sliced cheese. *Food Sci. Technol., 81*, 314–318.
3. Fang, Z., Zhao, Y., Warner, R., & Johnson, S. K., (2017). Active and intelligent packaging in meat industry. *Trends Food Sci Technol., 61* 60–71.
4. Realini, C. E., & Marcos, B., (2014). Active and intelligent packaging systems for a modern society. *Meat Science, 98*(3), 404–419.
5. Atarés, L., & Chiralt, A., (2016). Essential oils as additives in biodegradable films and coatings for active food packaging. *Trends Food Sci. Technol., 48*, 51–62.
6. Ribeiro-Santos, R., Andrade, M., Ramos de Melo, N., & Sanches-Silva, A., (2017). Use of essential oils in active food packaging: Recent advances and future trends. *Trends Food Sci. Technol., 61*, 132–140.
7. Chávez-González, M. L., Rodríguez-Herrera, R., & Aguilar, C. N., (2016). Essential oils: A natural alternative to combat antibiotics resistance. *Antibiotic Resistance: Mechanism and New Antimicrobial Approaches*. Academic Press, Elsevier, 227–237.
8. Rivera, C. K., Crandall, P. G., O'Bryanm C. A., & Ricke, S., (2015). Essential oils as antimicrobials in food systems- a review. *Food Control, 54*, 111–119.
9. Bakkali, F., Averbeck, S., Averbeck, D., & Idaomar, M., (2008). Biological effects of essential oils-a review. *Food Chem. Toxicol., 46*, 446–75.
10. Burt, S., (2004). Essential oils: Their antibacterial properties and potential applications in foods: a review. *Int. J. Food Microbiol., 94*(3), 223–253. http://dx.doi.org/10.1016/j.ijfoodmicro.2004.03.022.
11. Pichersky, E., Noel, J. P., & Dudareva, N., (2006). Biosynthesis of plant volatiles: Nature's diversity and ingenuity. *Science, 311*, 808–811.
12. Bomfim, N. S., Nakassugi, L. P., Oliveira, J. F. P., Kohiyama, C. Y., Mossini, S. A. G., GrespaN, R., et al., (2015). Antifungal activity and inhibition of fumonisin production by *Rosmarinus officinalis* L. essential oil in *Fusarium verticillioides* (Sacc.) Nirenberg. *Food Chem., 166*, 330–336.

13. Riahi, L., Chograni, H., Elferchichi, M., Zaouali, Y., Zoghlami, N., & Mliki, A., (2013). Variations in Tunisian wormwood essential oil profiles and phenolic contents between leaves and flowers and their effects on antioxidant activities. *Ind. Crop Prod.*, *46*, 290–296. http://dx.doi.org/10.1016/j.indcrop.2013.01.036
14. Sánchez-González, L., Vargas, M., González-Martínez, C., Chiralt, A., & Cháfer, M., (2011). Use of essential oils in bioactive edible coatings: A review. *Food Eng. Rev.*, *3*(1), 1–16.
15. Wen, P., Zhu, D. H., Wu, H., Zong, M. H., Jing, Y. R., & Han, S. Y., (2016). Encapsulation of cinnamon essential oil in electrospun nanofibrous film for active food packaging. *Food Control*, *59*, 366–376.
16. Krepker, M., Shemesh, R., Poleg, Y. D., Kashi, Y., Vaxman, A., & Segal, E., (2017). Active food packaging films with synergistic antimicrobial activity. *Food Control*, *76*, 117–126.
17. Cha, D. S., & Chinnan, M., S., (2004). Biopolymer–based antimicrobial packaging: A review. *Crit. Rev. Food Sci. Nutr.*, *44*(4), 223–237. http://dx.doi.org/10.1080/10408690490464276.
18. Paduch, R., Kadefer-Szerszen, M., Trytek, M., & Fiedurek, J., (2007). Terpenes: Substances useful in human healthcare. *Arch. Immunol. Ther. Exp.*, *55*, 315–327.
19. Helander, I. M., Alakomi, H. L., Latva-Kala, K., Mattila-Sandholm, T., Pol, L., Mid, E. J., et al., (1998). Characterization of the action of selected essential oil components on gram-negative bacteria. *J. Agric. Food Chem.*, *46*, 3590–3595.
20. Calo, J. R., Crandall, P. G., O'Bryan, C. A., Ricke, S. C., (2015). Essential oils as antimicrobials in food systems-A Review. *Food Control*, *54*, 111–119.
21. Hashemi, S. M. G., & Khaneghah, A. M., (2017). Characterization of novel basil-seed gum active edible films and coatings containing oregano essential oil. *Prog. Org. Coat.*, *110*, 35–41.
22. Andrade, K. S., Poncelet, D., & Ferreira, S. R. S., (2017). Sustainable extraction and encapsulation of pink pepper oil. *J. Food Eng.*, *204*, 38–45.
23. Tekin, K., Akalin, M. K., & Seker, M. G., (2015). Ultrasound bath-assisted extraction of essential oils from clove using central composite design. *Ind. Crops Prod.*, *77*, 954–960.
24. Shahbazi Y., (2017). The properties of chitosan and gelatin films incorporated with ethanolic red grape seed extract and *Ziziphora clinopodioides* essential oil as biodegradable materials for active food packaging. *Int. J. Biol. Macromolec.*, *99*, 746–753.
25. Martucci, J. F., Gende, L. B., Neira, L. M., & Ruseckalite, R. A., (2015). Oregano and lavender essential oils as an antioxidant and antimicrobial additives of biogenic gelatin films. *Ind. Crops Prod.*, *71*, 205–213.
26. El Sayed, H., Chizzola, R., Ramadan, A. A., & Edris, A. E., (2017). Chemical composition and antimicrobial activity of garlic essential oils evaluated in organic solvent, emulsifying, and self-microemulsifying water based delivery systems. *Food Chem.*, *221*, 196–204.
27. Baba, E., Acar, Ü., Öntas, C., Kesbic, O., & Yilmaz, S., (2016). Evaluation of citrus limone peles essential oil on growth performance, immune response of Mozambique tilapa *Oreochromis mossambicus* challenged with *Edwardsiella tarda*. *Aquaculture*, *465*, 13–18.

28. Verma, R. S., Padalia, R. C., Goswami, P., Upadhyay, R. K., Singh, V. R., Chauhan, A., et al., (2016). Assessing productivity and essential oil quality of Himalayan thyme (*Thymus linearis* Benth). In the subtropical region of north India. *Ind. Crops Prod.*, *94*, 557–561.
29. Elansary, H. O., Yessoufou, K., Shokralla, S., Mahmoud, E. A., & Wozniak, K. S., (2016). Enhancing min and basil oil composition and bacteria activity using seaweed extracts. *Ind. Crops Prod.*, *92*, 50–56.
30. Van Haute, S., Raes, K., Devlieghere, F., & Sampers, I., (2017). Combined use of cinnamon essential oil and MAP/vacuum packaging to increase the microbial and sensorial shelf life of lean pork and salmon. *Food Packaging and Shelf Life*, *12*, 51–58.
31. Mateo. E. M., Gómez, J. V., Domínguez, I., Gimeno-Adelantado, J. V., Mateo-Castro, R., Gavara, R., et al., (2017). Impact of bioactive packaging systems based on EVOH films and essential oils in the control of aflatoxigenic fungi and aflatoxin production in maize. *Int. J. Food Microbiol.*, *254*, 36–46.
32. Gherardi, R., Becerril, R., Nerin, C., & Bosetti, O., (2016). Development of a multilayer antimicrobial packaging material for tomato puree using an innovative technology. *Food Sci. Technol.*, *72*, 361–367.
33. Han, C., Wang, J., Li, Y., Lu, F., & Cui, Y., (2014). Antimicrobial-coated polypropylene films with polyvinyl alcohol in packaging of fresh beef. *Meat Science*, *96*, 901–907.
34. Kwon, S. J., Chang, Y., & Han, J., (2017). Oregano essential oil-based natural antimicrobial packaging film to inactivate *Salmonella enterica* and yeasts/molds in the atmosphere surrounding cherry tomatoes. *Food Microbiol.*, *65*, 114–121.
35. Paparella, A., Mazzarrino, G., Chaves-López, C., Rossi, C., Sacchetti, G., Guerrieri, O., et al., (2016). Chitosan boosts the antimicrobial activity of *Origanum vulgare* essential oil in modified atmosphere packaged pork. *Food Microbiol.*, *59*, 23–31.
36. Muriel-Galet, V., Cerisuelo, J. P., López Carballo, G., Aucejo, S., Gavara, R., & Hernández-Muñoz, P., (2013). Evaluation of EVOH-coated PP films with oregano essential oil and citral to improve the shelf life of packaged salad. *Food Control*, *30*, 137–143.
37. Hulankova, R., Borilova, G., & Steinhauserova, I., (2013). Combined antimicrobial effect of oregano essential oil and caprylic acid in miced beef. *Meat Science*, *95*, 190–194.
38. Sirocchi, V., Devlieghere, F., Peelman, N., Sagratini, G., Maggi, F., Vittori, S., et al., (2017). Effect of *Rosmarius officinlis* L. essential oil combined with different packaging conditions to extend the shelf life of refrigerated beef meat. *Food Chem.*, *221*, 1069–1076.
39. Sangsuwan, J., Pongsapakworawat, T., Bangmo, P., & Sutthasupa, S., (2016). Effect of chitosan beads incorporated with lavender or red thyme essential oils in inhibiting *Botrytis cinerea* and their application in strawberry packaging system. *Food Sci. Technol.*, *74*, 14–20.
40. Padula, M., & Ito, D., (2006). Food packaging and safety. In: *Bulletin of Technology and Development of Packaging*, *18* (3), 1–6.

41. Mulla, M., Ahmed, J., Al-Attar, H., Castro-Aguirre, E., Arfat, Y. A., & Auras, R., (2017). Antimicrobial efficacy of clove essential oil infused into chemically modified LLDPE film for chicken meat packaging. *Food Control, 73*, 663–671.
42. Sung, S. Y., Sin, L. T., Tee, T. T., Bee, S. T., Rahmat, A. R., & Rahman, W. A. W. A., (2014). Control of bacteria growth on ready-to-eat beef loaves by antimicrobial plastic packaging incorporated with garlic oil. *Food Control, 39*, 214–221.
43. Atarés, L., & Chiralt, A., (2016). Essential oils as additives in biodegradable films and coatings for active food packaging. *Trends Food Sci. Technol., 48*, 51–62.
44. Shojaee-Aliabadi, S., Hosseini, H., Mohammadifar, M. A., Mohammadi, A., Ghasemlou, M., Ojagh, S. M., et al., (2013). Characterization of antioxidant antimicrobial K-carrageenan films containing *Satureja hortensis* essential oil. *Int. J. Biol. Macromolec., 52*(1), 116e124. http://dx.doi.org/10.1016/j.ijbiomac.2012.08.026.
45. Salgado, P. R., López-Caballero, M. E., Gómez-Guillén, M. C., Mauri, A. N., & Montero, M. P., (2013). Sunflower protein films incorporated with clove essential oil have potential application for the preservation of fish patties. *Food Hydrocoll., 33*(1), 74–84.
46. Galus, S., & Kadzińska, J., (2016). Whey protein edible films modified with almond and walnut oils. *Food Hydrocoll., 52*, 78–86.
47. Klangmuang, P., & Sothornvit, R., (2016). Barrier properties, mechanical properties and antimicrobial activity of hydroxypropyl methylcellulose-based nanocomposite films incorporated with Thai essential oils. *Food Hydrocoll., 61*, 609–616.
48. Moradi, M., Tajik, H., Razavi, R. S. M., & Mahmoudian, A., (2016). Antioxidant and antimicrobial effects of zein edible film impregnated with *Zataria multiflora boiss*. Essential oil and monolaurin. *LWT – Food Sci. Technol., 72*, 37–43.
49. Perdones, Á., Chiralt, A., & Vargas, M., (2016). Properties of film-forming dispersions and films based on chitosan containing basil or thyme essential oil. *Food Hydrocoll., 57*, 271–279.
50. Choi, W. S., Singh, S., & Lee, Y., S., (2016). Characterization of an edible film containing essential oils in hydroxypropyl methylcellulose and its effect on quality attributes of "Formosa" plum (*Prunus Salicina* L.). *LWT – Food Sci. Technol., 70*, 213–222.
51. Nur Fatin Nazurah, R., & Nur Hanani, Z. A., (2017). Physicochemical characterization of Kappa-Carrageenan (*Euchema Cottoni*) based films incorporated with various plant oils. *Carbohydr. Polym., 157*, 1479–1487.
52. Riquelme, N., Herrera, M. L., & Matiacevich, S., (2017). Active films based on alginate containing lemongrass essential oil encapsulated: Effect of process and storage conditions. *Food Bioprod. Process, 104*, 94–103.
53. Frazao, G. G. S., Blank, A. F., & De Aquino Santana, L. C. L., (2017). Optimization of an edible chitosan coatings formulations incorporating myrcia ovata cambessedes essential oil with antimicrobial potential against foodborne bacteria and natural microflora of mangaba fruits. *LWT – Food Sci. Technol., 79*, 1–10.
54. Acosta, S., Chiralt, A., Santamarina, P., Rosello, J., González-Martínez, C., & Cháfer, M., (2016). Antifungal films based on starch-gelatin blend, containing essential oils. *Food Hydrocoll., 61*, 233–240.

55. Teixeira, B., Marques, A., Pires, C., Ramos, C., Batista, I., Saraiva, J. A., et al., (2014). Characterization of fish protein films incorporated with essential oils of clove, garlic and origanum: Physical, antioxidant and antibacterial properties. *LWT-Food Sci. Technol.*, *59*, 533–539.
56. Sivarooban, T., Hettiarachchy, N. S., & Johnson, M. G., (2008). Physical and antimicrobial properties of grape seed extract, nisin, and EDTA incorporated soy protein edible films. *Food Research International*, *41*, 781–785.
57. Hashemi Gahruie, H., Ziaee, E., Eskandari, M. H., & Hosseini, S. M. H., (2017). Characterization of basil seed gum-based edible films incorporated with zataria multiflora essential oil nanoemulsion. *Carbohydr. Polym.*, *166*, 93–103.
58. Acevedo-Fani, A., Salvia-Trujillo, L., Rojas-Graü, M. A., & Martín-Belloso, O., (2015). Edible films from essential-oil-loaded nanoemulsions: Physicochemical characterization and antimicrobial properties. *Food Hydrocoll.*, *47*, 168–177.
59. Munhuweyi, K., Caleb, O. J., Lennox, C. L., Van Reenen, A. J., & Opara, U. L., (2017). In vitro and *in vivo* antifungal activity of chitosan-essential oils against pomegranate fruit pathogens. *Postharvest Biol. Technol.*, *129*, 9–22.
60. Sánchez, A. D., Andrade-ochoa, S., Aguilar, C. N., Contreras-esquivel, J. C., & Nevárez-Moorillón, G. V., (2015). Antibacterial activity of pectic-based edible films incorporated with Mexican lime essential oil. *Food Control*, *50*, 907–912.
61. Martucci, J. F., Gende, L. B., Neira, L. M., & Ruseckaite, R. A., (2015). Oregano and lavender essential oils as an antioxidant and antimicrobial additives of biogenic gelatin films. *Ind. Crops Prod.*, *71*, 205–213.
62. Lee, J. H., Lee, J., & Song, K., (2015). Bin. Development of a chicken feet protein film containing essential oils. *Food Hydrocoll.*, *46*, 208–215.
63. Seydim, A. C., & Sarikus, G., (2006). Antimicrobial activity of whey protein based edible films incorporated with oregano, rosemary and garlic essential oils. *Food Res. Int.*, *39*(5), 639–644.
64. Hafsa, J., Smach, M., Ali Ben Khedher, M. R., Charfeddine, B., Limem, K., Majdoub, H., & Rouatbi, S., (2016). Physical, antioxidant and antimicrobial properties of chitosan films containing eucalyptus globulus essential oil. *LWT – Food Sci. Technol.*, *68*, 356–364.
65. Pires, C., Ramos, C., Teixeira, B., Batista, I., Nunes, M. L., & Marques, A., (2013). Hake proteins edible films incorporated with essential oils: Physical, mechanical, antioxidant and antibacterial properties. *Food Hydrocoll.*, *30*(1), 224–231.

CHAPTER 7

PERSPECTIVES FOR FOOD DEVELOPMENT FROM PITAYO *Stenocereus queretaroensis* (WEBER) BUXBAUM

GAYTÁN-ANDRADE JUAN JOSÉ,
CRISTÓBAL N. AGUILAR,
LÓPEZ-LÓPEZ LLUVIA ITZEL, COBOS-PUC LUIS ENRIQUE, and
SILVA-BELMARES SONIA YESENIA

School of Chemistry, University of Coahuila, Saltillo, Coahuila, Mexico, E-mail: josegaytan@uadec.edu.mx

ABSTRACT

At present, the trend in the consumption of functional foods is increasing; the world population is interested in acquiring them due to its nutritional and functional benefits. The plant *Stenocereus queretaroensis* belongs to the *Cactaceae* family and it is endemic to Mexico. *Stenocereus queretaroensis* is widely distributed in the states of Jalisco, Querétaro, Zacatecas, Colima, and Aguascalientes. The *Cactaceae* family has great expansion potential such as the prickle pear that is commercialized internationally, reason why there's a great interest on cultivating this plant on marketing, processing alternatives and new food development. As with most products, to develop a *S. queretaroensis* food it is necessary to establish a quality control system based on fruit selection that includes size and color. *S. queretaroensis* represents a perspective for the development of new functional foods, for it contains phytochemical compounds that contribute to health benefits. The objective of this review is to provide some perspectives focused on the development of functional foods formulated with active ingredients of the *S. queretaroensis* fruit.

7.1 INTRODUCTION

The *Stenocereus queretaroensis* plant belongs to the *Cactaceae* family and it is endemic to Mexico, it has a columnar morphology, it is cultivated on arid and semi-arid regions and it produces edible fruits [1, 2]. There are other species reported of *Stenocereus* in Canada and Argentina. Some known species of *Stenocereus* are pitalla, pitahaya, pitahalla, pitajalla or pitajaya [3]. There are a large number of species of the family *Cactaceae* with columnar morphology, these developed physiological and morphological adaptations that allowed them to survive in desert habitats [4, 5]. These adaptations, gave them the capacity to tolerance droughts [6], reason why their cultivation is promising in arid, desert and semi-desert environments [7]. The *Stenocereus* plants propagate from cuttings, and the roots develop two to four weeks after the cut. If the plant piece is acquired from a mature plant, the fruit will be produced annually during a production period of 30 to 50 years. The fruits can be eaten fresh, dried or processed when the juice is extracted. The cultivation of this *Cactaceae* is promising since from a herd, there could be a production of more than 1,100 plants [8] (Figures 7.1a and 7.1b).

7.1.1 PHYSICAL-CHEMICAL AND NUTRITIONAL CHARACTERISTICS

The fruit is ovoid, it shows a transverse diameter of 7 to 17 cm, it has a weight of 150 to 350 g, its shell is smooth and it has a color that varies from green to red. It has spines depending on the type and degree of ripening. The fruit pulp could be red, yellow, white or pink and the taste is slightly sweet [9] (Figure 7.2).

Bromatological studies show the nutritional importance of the fruits of the *Stenocereus* genus. However, *S. queretaroensis* does not have information recorded about its nutritional value. Also, *S. queretaroensis* fruits show nutritional properties similar to tropical-fruits but with higher content of sodium, potassium and vitamin C. Compared to other plants of the *Cactaceae* family, the *Stenocereus* pericarp lacks of spines, which facilitates the harvest and post-harvest. Additionally, its seeds are smaller than the prickly pear's fruit and can be swallowed easily. The fruit shows an extraordinary beauty and therefore it is used to make fruit ornaments.

FIGURE 7.1A Long-lived plant.

FIGURE 7.1B Young plant.

FIGURE 7.2 Fruits of the genus *Stenocereus queretaroensis* weber.

Table 7.1 shows the percentage of dry and wet weight of the whole fruit, pulp, peel, and seed of *Stenocereus griseus* and *Stenocereus stellatus* [10–12].

Table 7.2 shows the physico-chemical analysis of the sap of *S. griseus* and *S. stellatus*. Table 7.2 also shows pH, as well as contents of soluble solids, ascorbic acid, total sugars, and reducer's sugars. These results demonstrate that *S. queretaroensis* presents higher sugar content than other *Stenocereus* species.

7.1.2 SUSTAINABILITY

The *S. queretaroensis* fruits are important for the economy of rural communities in Mexico such as Autlan, Sayula, Amacueca, Zacoalco de Torres, Techaluta de Montenegro and Atoyac, localized at the southeast of Jalisco [13]. These fruits are part of the identity of the population of these communities; therefore, it represents a market alternative for different needs, compared to other fruits of the family *Cactaceae*.

The main problem that this fruit has on a commerce level is that price collapses slightly after 15 days of the harvest. The fruit is marketed in an area known as "the nine corners of Guadalajara city," between the

TABLE 7.1 Proximal Composition of *Stenocereus* Genus

Proximate composition (g/100g) (%)	Genus						
	S. griseus					*S. stellatus*	
	Fruit complete	Pulp	Shell		Seed	Pulp	
	Wb	Wb	Db	Db	Db	Wb	Db
Crude Protein	6.93	1.29	9.07–9.15	8.39	21.75	1.00	7.33
Crude Fiber	16.76	3.29	14.75–23.15	13.54	5.09	0.25	1.83
Ash	6.67	0.46	3.23	-	-	0.48	3.54
Crude Fat	1.00	0.12	0.70–0.84	0.64	1.69	2.28	2.07
Nitrogen free extract (Carbohydrate)	68.64	9.05	63.71	-	-	11.66	85.23
Moisture	84.13	85.79	0.0	0.0	0.0	86.33	0.0

-; Not Date, Wb; wet basis, Db; dry basis.

TABLE 7.2 Physicochemical Parameters of the Genus *Stenocereus*

Parameters	*S. griseus*	*S. stellatus*	*S. queretaroensis*				
			Yellow	White	Mamey	Purple	Red
Titratable acidity (citric acid) (g/100 mL)	0.115	0.640	0.50	0.18	0.15	0.29	0.17
pH	5.20	3.95	3.9	4.7	5.0	4.6	4.9
Dissolved solids to 20°C (%)	12.2	9.1	-	-	-	-	-
Solids in suspension (%)	4.0	0.7	-	-	-	-	-
Reducing sugar (%)	8.12	7.90	10	11	9	10	10
Total sugar (%)	8.61	8.10	11	11	10	11	10
Vitamin C (mg ascorbic acid/100 mL)	21.7	11.72	-	-	-	-	-

-; Not date.

Paz Avenue and Cristobal Colon Street, by producers from different communities.

The producers compete for quality and price, while the customers wait for the season to taste the fruits. The season for commercialization is limited to the months of April and May, because the beginning of the rains causes loss fruits. The fruit color commercialized is varied; they can present a color red, pink, yellow, melon, orange, purple and other shades.

In the state of Jalisco, some products are produced with the fruit of *S. queretaroensis* using a simple technology, because the fruit usually presents a useful life of three days after the cut.

Food preservation techniques are employed to prolong the fruit shelf life. The products are handcrafted; jelly is an example, whose shelf life lasts up to one year. In the southeast of Jalisco, more than 15 products displayed in informal stores, some them are presented in Figure 7.3.

7.1.3 OPPORTUNITY SECTORS

In Mexico, the areas to grow and harvest *S. queretaroensis* have increased considerably since wild growing areas under arid and semi-arid conditions are limited, as a result, the cultivation of S. *queretaroensis* fruits have an enormous potential for regional, national and international markets [15] This fruit has a great demand for its organoleptic properties, colors range,

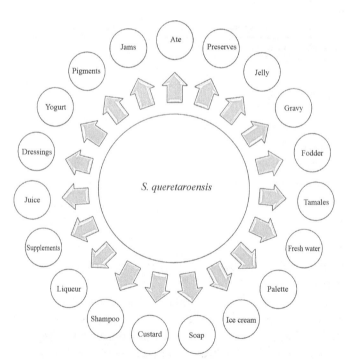

FIGURE 7.3 Products made with the genus *Stenocereus*.

flavors, and odors [16]. The fruit is very appreciated by the international market since it represents a source of income and employment. All of its components accordingly (fruit, peel, sap, and seed) have a potential for production and marketing [17].

The cultivation of this fruit is carried out more frequently in home gardens, but it is perishable and the transport routes for commercial distribution are inadequate. Researches on this fruit are scarce, so it can be a productive alternative [18], because it can be used as raw material for product development for food and cosmetic industries [19]. The Trade in Asia has increased, and the collection of wild fruits had been done in excess, so a balance between use and conservation is needed. This includes strategies implementation with ecological, technological and socio-economic aspects for each region. Accordingly, it is imperative to carry out research that allows, in the short term, to complete and systematize a diagnosis of plant genetic resources in Mexico to assess the current state of potentiation (Figure 7.4).

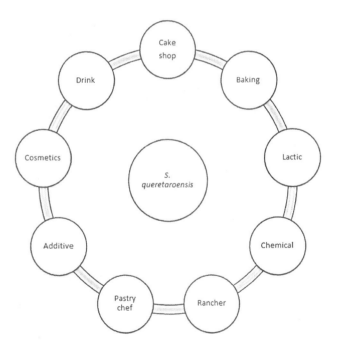

FIGURE 7.4 Industries interested in the development of products of the genus *Stenocereus*.

KEYWORDS

- food development
- perspectives
- pitaya fruit
- *Stenocereus queretaroensis*

REFERENCES

1. Pimienta, E., Pimienta, E., & Nobel, P., (2004). Ecophysiology of Querétaro pitayo *(Stenocereus queretaroensis). J. Arid. Environ., 59,* 1–17.
2. Rebollar, A., Romero, J., Cruz, P., & Zepeda, H., (2002). The cultivation of the pitaya *(Stenocereus* spp.), an alternative for the dry tropics of the state of Michoacan. Autonomous University of Chapingo. México, 2nd Edition.
3. Mercado, J., & Granados, D., (2007). The pitaya. Biology, Ecology, Physiology .Mizrahi, Y., Raveh, E., Yossov, E., Nerd, A., and Ben-Asher, J.
4. Bravo, H., (1978). *The Cactaceae of Mexico* (2nd edition, Vol. I, p. 743), UNAM.
5. Mauseth, J., (2000). Theoretical aspects of the surface to volume ratios and water storage capacities of succulent shoots. *Amer. J. Botany., 88,* 11071115.
6. Sowel, J., (2001). *Desert Ecology.* An introduction to life in the arid Southwest. University of Utah Press, Salt Lake City, USA, p. 192.
7. El-Obeidy, A., (2004). Introducing and growing some fruiting columnar cacti in a new arid environment, Department of Fruit Horticulture, Faculty of Agriculture Cairo University, Giza, EGYPT, *Journal of Fruit and Ornamental Plant Research, 12,* 127–136.
8. Casas, A., Pickersgill, B., Cabal lero, J., & Valient, B. A., (1997). Ethnobotany and domestication in xoconostle, *Stenocereus stellatus (Cactaceae),* in the Tehuacan Valley and La Mixteca Baja, Mexico. *Economic Botany, 51,* 279–292.
9. Flores, C., (2003). Pitayas and Pitahayas. Production, post-harvest, industrialization and commercialization. Autonomous University of Chapingo. México, 1st edition
10. Martínez, G.J., (1993). Pitaya *(Stenocereus griseus)* type characterization. Haworth at Mixteca. Thesis. UACH. México.
11. Mercado, J., & Granados, D., (2002). The pitaya. Biology, Ecology, Systematic Physiology, Ethnobotany. Autonomous University of Chapingo. México, 2nd edition.
12. Bravo, H. H., & Sánchez, M., (1991). The Cactaceae of Mexico. México. 3rd ed. Vol 1 & 3. UNAM México., p. 643.
13. Acosta, G. L., (2000). Microbial fermentation of pitaya pulp *(S. queretaroensis)* for the release of water-soluble pigment. Thesis UACH. México., p. 45.

14. López, G. R., Díaz, P. J., & Flores, M., (2000). Vegetative propagation of three species of cacti: Pitaya (*Stenocereus griseus*), Tunillo (*Stenocereus stellatus*) and Jiotilla (*Escontria chiotilla*). *Agrociencia.*, *34*, 363–367.
15. Piña, L. I., (1977). Pitayas and other related cacti of the state of Oaxaca. *Mexican Succulent Cactaceae, 22*(1), 3–13.
16. Pimienta, B. E., & Nobel, P. S., (1994). Pitaya (*Stenocereus* spp., Cactaceae): An ancient and modern fruit crop of México. *Economic Botany, 48,* 76–83.
17. Campos, R. E., Pinedo, E. J., Campos, M. R., & Hernandez, F. A., (2011). Evaluación de plantas de pitaya (*Stenocereus spp*) de poblaciones naturales de Monte Escobedo, Zacatecas. *Rev. Chapingo Ser.Hortic.*, *17*, p. 3.
18. Casas, A., (1998). Domesticación de plantas y recursos genéticos de México. Boletín de la Sociedad Botánica de México, *62*, 73–76.
19. Emaldy, U., & Semprum, C., (2006). Cardon dato (*Stenocereus griseus*) Cactaceae fruit pulp as raw material form marmalade production. *Latin American Nutrition Archives, 56* (1), 83–89.

CHAPTER 8

TEA FROM Camellia sinensis: A NEW TENDENCY OF VALUABLE ACTIVE COMPOUNDS

A. OCHOA-CHANTACA, G. C. G. MARTÍNEZ-ÁVILA,
E. J. SÁNCHEZ-ALEJO, and R. ROJAS

Autonomous University of Nuevo León, Research Center and Development for Food Industries (CIDIA). School of Agronomy, General Escobedo, Nuevo León, 66050, México, E-mail: romeo.rojasmln@uanl.edu.mx

ABSTRACT

The popular use of the medicinal plants is part of the cultural heritage of a village. This knowledge is transmitted from generation to generation without a record and some rural communities consider the traditional medicine as the main axis of their health system. *Camellia sinensis* is a plant from China, but now is harvested around the world and is one of the main herbs that is used in the traditional Chinese medicine. In the last years, the consumption of natural beverages is increased due to their phenolic composition and biological properties like antioxidant, anticancer activity, prevent the coronary heart diseases and is used for the treatment of different diseases like rheumatism, inflammation of respiratory and intestinal tract, gastric disturbance and venereal disease. Also, the presence or absence of some compounds is due to the type of tea, processing, date of harvest, extraction method and others factors that need of consider. In this chapter, an overview of the most recent advances about of green, white, black and red tea from *Camellia sinensis*. Their properties, activities, main phenolic compounds and extract methods are presented.

8.1 BACKGROUND

The consumption of bioactive compounds from natural sources in food new tendency due to their biological activities of these compounds, antioxidant potential that permit the production of food without synthetic antioxidants (butylated hydroxyanisole and butylated hydroxytoluene (BHT) of 1.0 and 0.25 mg/kg body weight/day – acceptable daily intake) for consumers [1, 2]. Daily drinks as teas are a source of compounds that allow the intake of doses that can help improve health and avoid the consumption of synthetic antioxidants avoiding the controversies that generate their use.

Tea (*Camellia sinensis*) is the most consumed drink in the world next to water and the amount of consumption well exceeds coffee, beer, wine, and soft drinks [3]. The reasons for its popularity is the unique aroma, flavor and recently due to its potential health benefits against several diseases mostly from antioxidant activities from flavonoids present in tea [4]. Also, humans are exposed to widespread form of toxins via direct or indirect pathways. Air, water, soil, foods, fruits, vegetables, and animals are the main exposure pathways. In this way, it is clear that there is a great scope for drug discovery from traditional medicines. In fact, these days plants have an important role in Western medicine as alternative sources of drugs for treating some diseases [5].

However, it is necessary to make several considerations for the presentation of tea's activities. From the cultivation method, place of cultivation, season of cultivation, form of cultivation, drying, transport, storage, extraction process, dose, frequency, form of intake and others. All these in order that the bioactive compounds are present in quantity and quality to exert the biological activities. In this review, we will give a summary of brief description of *Camellia sinensis*, extraction methods of main bioactive compounds using a traditional and emerging technologies, activities, uses and applications.

8.2 *CAMELLIA SINENSIS*

According to the Encyclopedia of genetics, Genomics, Proteomics and Informatics: *Camellia sinensis* are 82 species, 2n = 2x = 30. The epigallocatechin–3 component of green tea appears to be an angiogenesis inhibitor and thus can have anticancer effect [6]. Is an evergreen tree of the

THEACEAE family that grows to 10–15 m high in the wild, and 0.6–1.5 m under cultivation [7].

Catechin-(4-☐-8)-epi-gallocatechin. is originated from Southeast Asia and currently cultivated in more than 30 countries, including India, China, Sri Lanka, Kenya, Indonesia, Turkey, former Soviet Union, Japan, Iran, Bangladesh, Malawi, Vietnam and Argentina [8]. contain various components including polyphenols (36% of dry weight), methylxanthines (3.5%), amino acids (4%), organic acids (1.5%), carotenoids (<0.1%), volatiles (0.1%), carbohydrates (25%), protein (15%), lignin (6.5%), lipids (2%), ash (5%), chlorophyll and small amount of the other substances (0.5%), although the percentage of these ingredients could be changed by season, climate, horticultural practices and age of the leaf.

8.2.1 BLACK TEA

Black tea is a completely fermented tea [3, 9–11], for the preparation of this tea the fresh leaves are crushed and allowed to wither to induce oxidation as part of the tea fermentation process before drying [3], in the processing of this tea, the steaming or frying stages are omitted, but the leaves are ground to interrupt cell compartmentalization and bring the phenolic components in the leaves closer to the enzyme PPO (polyphenol oxidase) [12]. As a result, of fermentation (generally more than 80%) [3], a large proportion of the catechins are condensed into larger polyphenols known as theaflavins and thearubings which contribute to the characteristic reddish-black color, reduced astringency and bitterness, as well as elimination of the leafy and grassy flavor in black tea [3, 12].

8.2.2 WHITE TEA

White tea is an unfermented tea [3, 9, 11] and is made of young unopened bud leaves covered with small silver hairs, and the leaves are harvested once a year in early spring. Thus, sheets are then quickly vaporized and dried, with a minimal amount of processing to prevent oxidation [13], which gives it a light and delicate flavor [14]. Minimal processing is believed to result in the preservation of high amounts of phytochemicals conferring health benefits and their advantage over other types of tea [13].

8.2.3 RED TEA

Red tea is also known as Pu-ert, is a partially fermented tea [10], is obtained by a post-harvest fermentation stage before drying and vaporization, in which fermentation is achieved using microorganisms [15] and temperatures around 50°C [16]. The development of the special flavor of tea is attributed to the chemical changes of polyphenols, flavonols, caffeine, sugars, organic acids, amino acids and volatile flavor compounds during microbial fermentation [16].

8.2.4 GREEN TEA

It is an unfermented tea [3, 9, 11]. Green tea is generally made while the green leaf polyphenol ingredient, also known as catechin, does not begin to wither and to oxidize [17], are then subjected to the heat deactivation of the enzymes, the lamination and the final drying for dehydration; This results in minimal oxidation compared to the manufacture of other teas [18].

8.3 METHODS FOR EXTRACTION OF BIOACTIVE COMPOUNDS

8.3.1 CONVENTIONAL METHODS

Conventional extraction methods to obtain polyphenols from plants are commonly Soxhlet [19], maceration [20], hydro-distillation [21], infusion [22], decoction and rotary evaporator [23]. Soxhlet was invented by Franz von Soxhleth in 1879, represent the classical methodology for lipophilic compounds [24] and it is the main reference to which the performance of other extraction methods is compared [25]. This methodology has been used for more of one century to obtain different extracts types and is knowledge as the universal chemical extraction process [26]. Maceration is a conventional solid-liquid extraction that are mostly used for obtaining bioactive compounds from several materials like plants [27]. Also, is one method to produce increases in antioxidants in the process of wine production [28]. For steam extraction or hydro distillation, plant and water are put together into a flask and is heated until boiling. The steam takes the essential oil along and the water/oil mixture and is condensed to produce a two-phase system (water and essential oil), where the oil can be decanted and recovered. This method works due to the coexistence of two immiscible liquids [29]. Infusion is the process of constituent's extraction from plants material in solvents like water, edible oil or alcohol, by allowing the material to remain suspended in the solvent over time. Decoction is when the plant material and the solvent is put boiling, but aromas escape into the air [30] and finally, the rotary evaporator is most used to concentrate extracts (elimination of recovering of the solvent). For all these methods is needed the use of large quantities of solvent like ethanol, methanol, acetone, hexane and ethyl acetate [31, 32]. However, conventional extraction methods have some limitations such as high temperature, low efficiency, low yield, large quantity of solvents, mass transfer resistance, lead to degradation of thermally labile compounds, partial hydrolysis of water sensitive compounds [33] and health hazards [34]. These methods are currently commonly used to compare efficiency with more modern methods, but decoction is the method most used for the preparation of teas. In the next table shown the most recent research to obtain bioactive compounds from *Camellia sinensis* using conventional methods (Table 8.1).

8.3.2 EMERGING TECHNOLOGIES

Emerging technologies are another option to extract bioactive compounds from several materials. These extraction techniques are considered eco-friendly and associated with conserving resources [35]. Microwave [36], ultrasound [37], supercritical fluids [38, 39], high hydrostatic pressure [40] and pressurized hot water [41] are some of most used emerging technologies for polyphenols extraction. Besides, the quality of the extract depends on several factors like type of matrix nature, time, temperature, solvent, ratio, particle size, stirring, and others; but the emerging technologies commonly used only water, CO_2, ethanol or mixtures. These solvents have a low toxicity, present rapid mass transfer of the solutes and preserve the bioactivity of the compounds [39]. However, the selection of the method and solvent depend on the nature of compounds, easy of handling, level of the toxicity of process and the health hazard potential [42]. Finally, is needed to consider the geographical area of the plant, harvest season,

Table 8.1 Conventional Methods for Bioactive Compounds Extraction From Teas

Type of tea	Extraction method	Solvent	Conditions	Bioactive compounds	Ref.
Green	Maceration Rotary evaporator	EtOH (70%)	Room temperature (48 h), Storage (−20°C)	PH, FV, FVN.	[89]
White	Infusion (HW, CW)	Water	CW: 20–25°C/2 h HW: 70°C/7 min	FV, FVN, PH, TF, CT	[90]
Black	Infusion	Water	3, 6, 10 min/80 and 100°C	GA, ECG, 4-p-CQA, Q-3-OG-R-G, K-3-O-G-R-G, TF, TB	[91]
Black	Maceration	SC, EtOH (95%)	0 and 10 min/4000 rpm	TF, TH, CT	[92]
Green	Rotary evaporator	Water	100°C/15 min	E-3-G	[23]
Black, Green, Red, White	Infusion	Water	90°C/3 min	EGC, EC, EGCG, ECG	[93]
Green	Infusion	Water	Boiling water	CT, FV, HT	[94]
Black	Vortexed Centrifugal	EtOH	10 min/4000 rpm	TF, CT, TBS	[92]
Green	Cooking	Water Dimethyl ether	Water: 20°C/0.51 MPa DE:1.5 h and evaporation at 30°C	CT, CF	[95]
Black	Infusion	Water	100°C/30 min	CF, CT, GA	[96]
White Green Black Red	Infusion	Water	90°C/10 min	TN	[97]
Green	Maceration	EtOH MetOH	Room temperature/8 h	FV, FVN	[98]

Table 8.1 (Continued)

Type of tea	Extraction method	Solvent	Conditions	Bioactive compounds	Ref.
Green	Infusion	Water	HW/5 min	FVN, TN	[99]
Green	Maceration, Centrifugal	Water	43°C/1 h and centrifugation 4°C/10 min	PP	[100]
Green Black	Soxhlet	EtOH	500 mL EtOH/1 h	CT, TF, ETS, GFS	[71]

PH: Phenols, FV: Flavonols, FVN: Flavonoids, HW: Hot water, CW: Cold wáter, TF: Theaflavins, CT: Catechin, GA: Gallic acid, ECG: Epichatechinh-gallate, 4-p-CQA: 4-p-coumaroylquinc acid, Q-3-O-G-R-G: Quercetin-3-O-galactosyl-rhamnosyl-glucoside, K–3-O-G-R-G, Kaempferol-3-O-glucosyl-rhamnosyl-glucoside, TB: Theobromine, SC: Spodium chloride, TH: Thearubings, E-3-G: Epigallocatechin–3-gallato, HT: Hydrolyzable tannis, TBS: Thearobigins, EGC: Epigallocatechin, TN: Theanine, PP: Protein-polyphenol, ECG: Epichatechinh-gallate, EGCG: Epigallocatechim gallate, EC: Epicatechin.

used/studies morphological part, climate, ecological system, preparation and preservation of the vegetal material, type of drying, and others [43].

8.3.2.1 SUPERCRITICAL FLUIDS

Supercritical fluid extraction is an emerging technology that uses supercritical fluids (formed by one or more compounds at conditions over their critical values of pressure and temperature) as a solvent to separate the components from another (the matrix). In this process, the mobile phase is subjected to pressures and temperatures near or above the critical point for enhancing the mobile phase-solvating power [44]. The supercritical fluid is used as an alternative to traditional organic liquid solvents. The most widely used supercritical fluids are CO_2 (Tc = 31°C, Pc = 74 bar) and water (as above described) (Tc = 374°C, Pc = 221 bar), but some processes involve the use of methanol, ethanol, propane, ethane and others [45, 46].

8.3.2.2 MICROWAVE-ASSISTED EXTRACTION (MAE)

This technique is based in the heating principles of ionic conduction and dipole rotation of MAE. The electromagnetic energy is transferred rapidly on the bio-molecules by ionic conduction and dipole rotations, which result more power was dissipated inside the solvent and plant materials which generate molecular movement and heating on the extraction system rapidly and enhanced the loosening of the cell wall matrix quickly, increases the penetration of the solvent into the plant matrix which leads to enhanced the leaching of pectin from plant material to solvent [47, 48]. However, it has also been used for the extraction of bioactive compounds because it is one of the most promising emerging technology with advantages like fast extraction times with high yields, low cost, uses less solvent, and high-quality and stability of bioactive compounds compared with the conventional methods [49].

8.3.2.3 ULTRASOUND ASSISTED EXTRACTION

Ultrasound-assisted extraction is a nonconventional technique that involves mixing the sample with a solvent in a flask or baker and placing it in an ultrasonic bath, controlling the time, temperature and frequency [32]. Ultrasound-assisted extraction makes use of high-intensity ultrasonic energy created by the implosion of cavitation bubbles. The bubbles collapse can produce physical, chemical, and mechanical effects. When this energy reaches the surface of the raw material through the extraction solvent, it is transformed into mechanical energy that is equivalent to several thousand atmospheres of pressure [50]. The high-pressure breaks the material particles, destroys cell membranes, improves penetration of solvent, and increases the contact surface area between the solid and liquid faces resulting in the release of phenolic to the extraction solvent in a relative short time [50–52].

8.3.2.4 HIGH HYDROSTATIC PRESSURE

The high hydrostatic pressure method is a cold isostatic pressure processing method that can increase mass transfer rates and solvent permeability in the raw material, thereby leading to higher extraction yields and shorter extraction times [53, 55]. Has no negative effects on the activity and structure of bioactive compounds [56]. Also, consumes electric energy, does not cause exhaust emission and is recognized as an eco-friendly method by FDA [56, 57]. Besides, this method has been used in food preservation as an alternative to high-heat treatment, also, pressure of more than 300 MPa inactivates bacteria, yeast or viruses at room temperature within several minutes [58]. Some authors have been studied the extraction of bioactive compounds from *Camellia sinensis* improving the yields and stability of compounds [40, 59–60].

8.3.2.5 PRESSURIZED HOT WATER EXTRACTION (PHWE)

The first report of the use of PHWE was in 1994 reported by Hawthorne et al. [61], they reported the extraction of the organic pollutants from environmental solids and concluding that this method depends on the temperature, but present a little dependence of the pressure. PHWE is an extraction method that uses liquid water as extractant (extraction solvent) at temperatures above the atmospheric boiling point of water (100°C/273 K, 0.1 MPa), but below the critical point of water (374°C/647 K, 22.1 MPa). Also, can be referred as subcritical water extraction, superheated water extraction and pressurized liquid extraction using only water as a solvent [41]. The high temperature and pressure increases the solubility, diffusion rate and mass transfer of the compounds, decreases the viscosity and surface tension of the solvent [41, 62]. Besides, the extraction is faster than others conventional methodologies; but, the optimization process is needed due to the degradation of the bioactive compounds. This method has a highly energy-efficiency, eco-friendly (only use water as a solvent), is fast, present high yield and is a selective polyphenol extraction [63]. In the next table shown the most recent research to obtain bioactive compounds from *Camellia sinensis* using emerging technologies (Table 8.2).

8.4 MAIN BIOACTIVE COMPOUNDS

The composition of *Camellia sinensis* is very complex. However, some classes of compounds have been identified like terpenes, alkaloids (caffeine, theobromine, theophylline), flavanols (gallocatechin, catechin gallate, gallocatechin gallate, epicatechin, epigallocatechin, epicatechin gallate and epigallocatechin gallate) [64, 65], flavonols (kaempferol–3-O-(glucose-(1,3-rhamnose–1,6-glucose)), quercetin–3-O-(glucose-(1,3-rhamnose–1,6-galactose))) [66], phenolic acids (gallic acid, p-coumaric acid and quinic acid derivatives, caffeoylquinic acid isomers, and caffeoyl glucose) [67], volatile compounds, carotenoids, tocopherols, amino acids, polysaccharides (neutral(1 → 4)-□-galactan and pectin) [68, 69], vitamins (A, K, B, C), and minerals such as fluorine, potassium, magnesium, iron, manganese, and phosphorus, among others [70–72]. The major polyphenols belong to the family of catechins. This family comprises epigallocatechin gallate (EGCG), catechin (C), epicatechin (EC), gallocatechin (GC), gallocatechin gallate (GCG), epigallocatechin (EGC), and epicatechin gallate (ECG) [17] that contribute to antioxidant capacity and organoleptic properties [73]. The difference of polyphenols between non-fermented and partially or fully fermented teas have been reported, but, little information is available on the polyphenolic changes during fermentation [3].

Tea from Camellia sinensis 169

TABLE 8.2 Emerging Technologies for Bioactive Compounds Extraction From Teas

Type of tea	Extraction method	Solvent	Conditions	Bioactive compounds	Ref.
Green	SF	EtOH	43.7°C/106 min 19.3 MPa	(−)-Epigallocatechin−3-gallate (EGCG)	[101]
Green	PLE	EL, CO$_2$	10 MPa/100°C	CT	[102]
Green	MAE	Water	Boiled water: 0.5 min MAE: 1 min	CT, TN	[103]
Green	Infusión, UA	EtOH, Water	Infusion EtOH: 30 min Water: 15 min UAE: 26 min/24°C	FV, CT, EGC	[2]
Green	SF	EL, EA, EtOH	30 MPs/343°K	CT, CA, CF	[104]
Green	SF	EtOH	23MPa/63°C/120 min	CF, CT	[105]
Green	PLE	Water, EL, CO$_2$	373–473°K/20 min	CF, CT	[106]
Green	SF	CO$_2$	Decaffeinated: 250 bar/50°C Caffeine: 50, 60, 70, 80, 90°C/5, 10, 20, 40, 80 min/4°C	CF, CT	[107]
Green	MAE	Water	24,500 MHz/1000 W/80°C/10 min	CT	[108]
Green	MAE	MetOH	MAE: 55°C/3h MET: 80°C/ 3h	PLY, FVN	[109]
Green	MAE	Water	85°C/15 min	CF, TN, CT	[110]

CTT: Carmine tetrachloride, TCE: Tetrachloroethylene, SF: Supercritical fluid, CT: Catechin, ECG: Epichatechinh-gallate, EGCG: Epigallocatechim gallate, EGC: Epigallocatechin, EC: Epicatechin, PLE: Pressurized liquid extraction, EL: Ethyl lactate, MAE: Microwave-assisted extraction, TN: Theanine, EA: Ethyl acetate, CF: Caffeine, PLY: Polyphenols, FVN: Flavonoids, FV: Flavonols.

8.5 MAIN USES AND APPLICATIONS

The use of the center in traditional medicine. However, there are currently several studies that corroborate its effectiveness and identify which compounds are responsible for these properties. In most cases, it is due to the presence of the bioactive compounds, which obviously are quality, quantity, and conservation depend on many factors that go from the culture to the way of preparation of the tea, the methodology of extraction of the bioactive compounds and finally the dose, route of intake and frequency. The most important new uses and applications are related with the pharmaceutical industry such as antimicrobial [74, 75], antitumor [76], anti-cancer [77], anti-arthritic [78], anti-anaphylactic [79], anti-inflammatory [80], antimutagenic [81, 82], antioxidant [83, 84], antiproliferative [13, 85], gastrointestinal [71], hypoglycemic [86], hypolipidemic [86], anti-hyperglycemic [87], antidiabetic [88], and others. Is important to comment that the presence of ECGC might be the key factor in determining the neuroprotective role of green tea in brain diseases and cognitive performance [10].

8.6 SOLID STATE FERMENTATION AS EXTRACTION METHOD

Solid state fermentation (SSF) is a process in which microorganisms grow in an environment without free water or with very low free water content [113–115]. Its historical significance for the humanity dates back thousands of years, mainly to food processing, in both Western and Eastern countries. Taking into account the last century and the last decades, it is still used for the production of important biomolecules and products for many industries, such as food, pharmaceuticals, textiles, biochemicals and bioenergetics, among others [114]. In the last years has been used to the biotransformation of Teadenol A (Figure 8.1) from Pu-erh and quantify their antioxidant activity [111]. For this several microorganisms are used like *Aspergillus, Eurotium* and *A. tamari* using –2 g of dry weight, 60% aqueous EtOH, 40 min, room temperature-, –2 g dry weight, 60% aqueous EtOH (30 mL), room temperature/40 min-, –28°C/14 days-conditions, respectively. The biotransformation was from 0.23% to 1.79% of Teadenol A to 0.37% of Teadenol B (Figure 8.2) [111, 112]. Therefore, SSF

Tea from *Camellia sinensis* 171

is a viable strategy for obtaining compounds of high interest for various branches of the pharmaceutical and/or agro-food industry.

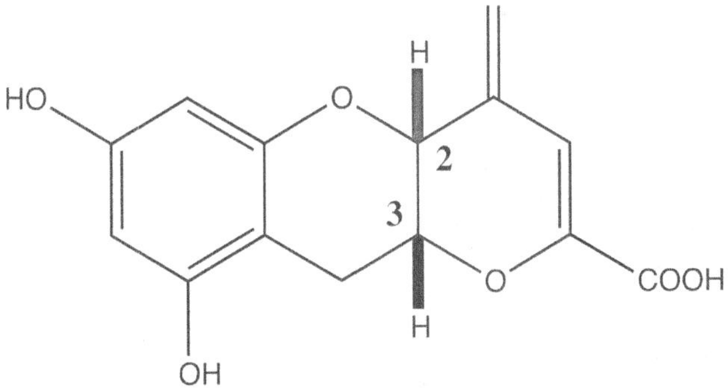

FIGURE 8.1 Teadenol A. (4a*R*,10a*R*)-7,9-dihydroxy-4-methylene-4,4a,10,10a-tetrahydropyrano[3,2-*b*]chromene–2-carboxylic acid.

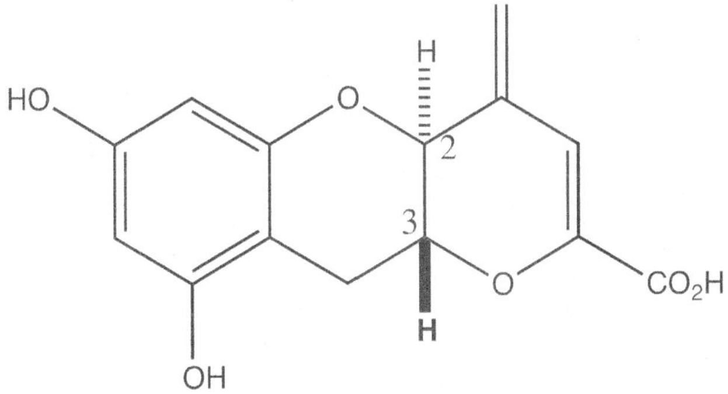

FIGURE 8.2 Teadenol B. (4a*S*,10a*R*)-7,9-dihydroxy-4-methylene-4,4a,10,10a-tetrahydropyrano[3,2-*b*]chromene–2-carboxylic acid.

8.7 CONCLUDING REMARK

The wide bioactive compounds content of tea is of great importance to health and has a protective effect against various diseases. However, it is necessary to evaluate the effect of these compounds on other diseases by inclusion in food. In addition, it is necessary to consider the extraction method extraction, form and type of cultivation, season, stability in food matrices, bioavailability, absorption and others, to be able to choose the best option without affecting the integrity of bioactive compounds. Also, it should be within the reach of all for be viable alternative of daily use, avoiding, reducing or substituting the use of synthetic compounds.

KEYWORDS

- bioactive compounds
- polyphenols
- tea

REFERENCES

1. Ares, G., Giménez, A., & Deliza, R., (2010). Influence of three non-sensory factors on consumer choice of functional yogurts over regular ones. *Food Quality and Preference, 21*(4), 361–367.
2. Lorenzo, J. M., & Munekata, P. E. S., (2016). Phenolic compounds of green tea: Health benefits and technological application in food. *Asian Pacific Journal of Tropical Biomedicine, 6*(8), 709–719.
3. Kim, Y., Goodner, K. L., Park, J. D., Choi, J., & Talcott, S. T., (2011). Changes in antioxidant phytochemicals and volatile composition of *Camellia sinensis* by oxidation during tea fermentation. *Food Chemistry, 129*(4), 1331–1342.
4. Cheng, T. O. (2006). All teas are not created equal. *International Journal of Cardiology, 108*(3), 301–308.
5. Phillipson, J. D., & Anderson, L. A., (1989). Ethnopharmacology and western medicine. *Journal of Ethnopharmacology, 25*(1), 61–72.
6. Rédei, G.P. Thea (*Camellia sinensis*), (2008). In: *Encyclopedia of Genetics, Genomics, Proteomics and Informatics*, Springer Netherlands: Dordrecht, pp. 1959–1959.
7. Ross, I.A. *Camellia sinensis*, (2005). In: *Medicinal Plants of the World: Chemical Constituents, Traditional and Modern Medicinal Uses* (Vol. 3). Humana Press: Totowa, NJ, pp. 1–27.
8. Graham, H. N., (1992). Green tea composition, consumption, and polyphenol chemistry. *Preventive Medicine, 21*(3), 334–350.
9. Yener, S., Sánchez-López, J. A., Granitto, P. M., Cappellin, L., Märk, T. D., Zimmermann, R., et al., (2016). Rapid and direct volatile compound profiling of black and green teas (*Camellia sinensis*) from different countries with PTR-ToF-MS. *Talanta., 152*, 45–53.
10. Schimidt, H. L., Garcia, A., Martins, A., Mello-Carpes, P. B., & Carpes, F. P. Green tea supplementation produces better neuroprotective effects than red and black tea in Alzheimer-like rat model. *Food Research International. 100* (1), 442–448.

11. Scoparo, C. T., Souza, L. M., Dartora, N., Sassaki, G. L., Santana-Filho, A. P., Werner, M. F. P., et al., (2016). Chemical characterization of heteropolysaccharides from green and black teas (*Camellia sinensis*) and their anti-ulcer effect. *International Journal of Biological Macromolecules, 86*, 772–781.
12. Muniandy, P., Shori, A. B., & Baba, A. S., (2017). Comparison of the effect of green, white and black tea on Streptococcus thermophilus and Lactobacillus spp. in yogurt during refrigerated storage. *Journal of the Association of Arab Universities for Basic and Applied Sciences, 22*, 26–30.
13. Hajiaghaalipour, F., Kanthimathi, M. S., Sanusi, J., & Rajarajeswaran, J., (2015). White tea (*Camellia sinensis*) inhibits proliferation of the colon cancer cell line, HT-29, activates caspases and protects DNA of normal cells against oxidative damage. *Food Chemistry, 169*, 401–410.
14. Espinosa, C., López-Jiménez, J. Á., Cabrera, L., Larqué, E., Almajano, M. P., Arnao, M. B., et al., (2012). Protective effect of white tea extract against acute oxidative injury caused by adriamycin in different tissues. *Food Chemistry, 134* (4), 1780–1785.
15. Soares, M. B., Izaguirry, A. P., Vargas, L. M., Mendez, A. S. L., Spiazzi, C. C., et al., (2013). Catechins are not major components responsible for the beneficial effect of *Camellia sinensis* on the ovarian δ-ALA-D activity inhibited by cadmium. *Food and Chemical Toxicology, 55*, 463–469.
16. Zhu, Y., Luo, Y., Wang, P., Zhao, M., Li, L., Hu, X., & Chen, F., (2016). Simultaneous determination of free amino acids in Pu-erh tea and their changes during fermentation. *Food Chemistry, 194*, 643–649.
17. Rameshrad, M., Razavi, B. M., & Hosseinzadeh, H., (2017). Protective effects of green tea and its main constituents against natural and chemical toxins: A comprehensive review. *Food and Chemical Toxicology, 100*, 115–137.
18. Han, Z. X., Rana, M. M., Liu, G. F., Gao, M. J., Li, D. X., Wu, F. G., et al., (2016). Green tea flavor determinants and their changes over manufacturing processes. *Food Chemistry, 212*, 739–748.
19. Pereira, M. G., Hamerski, F., Andrade, E. F., Scheer, A., D. P., & Corazza, M. L., (2017). Assessment of subcritical propane, ultrasound-assisted and Soxhlet extraction of oil from sweet passion fruit (Passiflora alata Curtis) seeds. *The Journal of Supercritical Fluids. 128*, 338–348.
20. Deng, J., Xu, Z., Xiang, C., Liu, J., Zhou, L., Li, T., et al., (2017). Comparative evaluation of maceration and ultrasonic-assisted extraction of phenolic compounds from fresh olives. *Ultrasonics Sonochemistry, 37*, 328–334.
21. Samadi, M., Abidin, Z. Z., Yunus, R., Awang Biak, D. R., Yoshida, H., & Lok, E. H., (2017). Assessing the kinetic model of hydro-distillation and chemical composition of Aquilaria malaccensis leaves essential oil. *Chinese Journal of Chemical Engineering, 25*(2), 216–222.
22. Fotakis, C., Tsigrimani, D., Tsiaka, T., Lantzouraki, D. Z., Strati, I. F., Makris, C., et al., (2016). Metabolic and antioxidant profiles of herbal infusions and decoctions. *Food Chemistry, 211*, 963–971.
23. Mehdipour, M., Daghigh Kia, H., Najafi, A., Vaseghi Dodaran, H., & García-Álvarez, O., (2016). Effect of green tea (*Camellia sinensis*) extract and pre-freezing equili-

bration time on the post-thawing quality of ram semen cryopreserved in a soybean lecithin-based extender. *Cryobiology, 73*(3), 297–303.
24. Cicero, A. M., Pietrantonio, E., Romanelli, G., & Di Muccio, A., (2000). Comparison of Soxhlet, shaking, and microwave-assisted extraction techniques for determination of PCB congeners in a marine sediment. *Bulletin of Environmental Contamination and Toxicology, 65*(3), 307–313.
25. Subramanian, R., Subbramaniyan, P., Noorul Ameen, J., & Raj, V., (2016). Double bypasses Soxhlet apparatus for extraction of piperine from Piper nigrum. *Arabian Journal of Chemistry, 9*, 537–540.
26. Heleno, S. A., Diz, P., Prieto, M. A., Barros, L., Rodrigues, A., Barreiro, M. F., et al., (2016). Optimization of ultrasound-assisted extraction to obtain mycosterols from Agaricus bisporus L. by response surface methodology and comparison with conventional Soxhlet extraction. *Food Chemistry, 197*, 1054–1063.
27. Belova, V. V., Voshkin, A. A., Kholkin, A. I., & Payrtman, A. K., (2009). Solvent extraction of some lanthanides from chloride and nitrate solutions by binary extractants. *Hydrometallurgy, 97*(3), 198–203.
28. Olejar, K. J., Fedrizzi, B., & Kilmartin, P. A., (2016). Enhancement of chardonnay antioxidant activity and sensory perception through maceration technique. *LWT – Food Science and Technology, 65*, 152–157.
29. Wollinger, A., Perrin, É., Chahboun, J., Jeannot, V., Touraud, D., & Kunz, W., (2016). Antioxidant activity of hydro distillation water residues from Rosmarinus officinalis L. leaves determined by DPPH assays. *Comptes Rendus Chimie, 19*(6), 754–765.
30. Visht, S., & Chaturvedi, S., (2012). Isolation of natural products. *Current Pharma Research, 2*(3), 584–599.
31. Ross, K. A., Beta, T., & Arntfield, S. D., (2009). A comparative study on the phenolic acids identified and quantified in dry beans using HPLC as affected by different extraction and hydrolysis methods. *Food Chemistry, 113*(1), 336–344.
32. Safdar, M. N., Kausar, T., Jabbar, S., Mumtaz, A., Ahad, K., & Saddozai, A. A., (2017). Extraction and quantification of polyphenols from kinnow (*Citrus reticulate* L.) peel using ultrasound and maceration techniques. *Journal of Food and Drug Analysis, 25*(3), 488–500.
33. Zhao, S., & Zhang, D., (2014). Supercritical CO_2 extraction of eucalyptus leaves oil and comparison with Soxhlet extraction and hydro-distillation methods. *Separation and Purification Technology, 133*, 443–451.
34. Jadhav, D., B.N, R., Gogate, P. R., & Rathod, V. K., (2009). Extraction of vanillin from vanilla pods: A comparison study of conventional Soxhlet and ultrasound assisted extraction. *Journal of Food Engineering, 93*(4), 421–426.
35. Yang, Y. C., & Wei, M. C., (2016). A combined procedure of ultrasound-assisted and supercritical carbon dioxide for extraction and quantitation oleanolic and ursolic acids from Hedyotis corymbosa. *Industrial Crops and Products, 79*, 7–17.
36. Calinescu, I., Lavric, V., Asofiei, I., Gavrila, A. I., Trifan, A., Ighigeanu, D., et al., (2017). Microwave-assisted extraction of polyphenols using a coaxial antenna and a cooling system. *Chemical Engineering and Processing: Process Intensification, 122*, 373–379.

37. Pradal, D., Vauchel, P., Decossin, S., Dhulster, P., & Dimitrov, K., (2016). Kinetics of ultrasound-assisted extraction of antioxidant polyphenols from food by-products: Extraction and energy consumption optimization. *Ultrasonics Sonochemistry, 32*, 137–146.
38. Da Porto, C., & Natolino, A., (2017). Supercritical fluid extraction of polyphenols from grape seed (Vitis vinifera): Study on process variables and kinetics. *The Journal of Supercritical Fluids, 130*, 239–245.
39. Da Silva, R. P. F. F., Rocha-Santos, T. A. P., & Duarte, A. C., (2016). Supercritical fluid extraction of bioactive compounds. *TrAC Trends in Analytical Chemistry, 76*, 40–51.
40. Xi, J., Shen, D., Zhao, S., Lu, B., Li, Y., & Zhang, R., (2009). Characterization of polyphenols from green tea leaves using a high hydrostatic pressure extraction. *International Journal of Pharmaceutics, 382*(1), 139–143.
41. Plaza, M., & Turner, C., (2015). Pressurized hot water extraction of bioactives. *TrAC Trends in Analytical Chemistry, 71*, 39–54.
42. Pandey, A., & Tripathi, S., (2014). Concept of standardization, extraction and pre phytochemical screening strategies for herbal drug. *J. Pharmacogn. Phytochem, 2*, 115–119.
43. Molyneux, R., & Colegate, S., (2007). An introduction and overview. In: *Bioactive Natural Products, CRC Press*, pp. 1–9.
44. Sairam, P., Ghosh, S., Jena, S., Rao, K. N. V., & Banji, D., (2012). Supercritical fluid extraction (SFE) – An overview. *Asian Journal of Research in Pharmaceutical Sciences, 2*(3), 112–120.
45. Zougagh, M., Valcárcel, M., & Ríos, A., (2004). Supercritical fluid extraction: A critical review of its analytical usefulness. *TrAC Trends in Analytical Chemistry, 23*(5), 399–405.
46. Ibañez, E., Herrero, M., Mendiola, J. A., & Castro-Puyana, M., (2012). Extraction and characterization of bioactive compounds with health benefits from marine resources: Macro and micro algae, cyanobacteria, and invertebrates. In: Hayes, M., (Ed.), *Marine Bioactive Compounds: Sources, Characterization and Applications* (pp 55–98). Springer US: Boston, MA.
47. Maran, J. P., Swathi, K., Jeevitha, P., Jayalakshmi, J., & Ashvini, G., (2015). Microwave-assisted extraction of pectic polysaccharide from waste mango peel. *Carbohydrate Polymers, 123*, 67–71.
48. Rojas, R., Contreras-Esquivel, J. C., Orozco-Esquivel, M. T., Muñoz, C., Aguirre-Joya, J. A., & Aguilar, C. N., (2015). Mango peel as source of antioxidants and pectin: Microwave-assisted extraction. *Waste Biomass Valor, 6*(6), 1095–1102.
49. Zhang, Y., Zheng, B., Tian, Y., & Huang, S., (2012). Microwave-assisted extraction and anti-oxidation activity of polyphenols from lotus (Nelumbo nucifera Gaertn.) seeds. *Food Science and Biotechnology, 21*(6), 1577–1584.
50. Virot, M., Tomao, V., Le Bourvellec, C., Renard, C. M. C. G., & Chemat, F., (2010). Towards the industrial production of antioxidants from food processing by-products with ultrasound-assisted extraction. *Ultrasonics Sonochemistry, 17*(6), 1066–1074.
51. Chemat, F., Vian, M. A., & Cravotto, G., (2012). Green extraction of natural products: Concept and principles. *International Journal of Molecular Sciences, 13*(7), 8615.

52. Castro-López, C., Rojas, R., Sánchez-Alejo, E. J., Niño-Medina, G., & Martínez-Ávila, G. C. G., (2016). Phenolic compounds recovery from grape fruit and by- products: An overview of extraction methods. In: Morata, A., & Loira, I., (Eds.), *Grape and Wine Biotechnology* (p. 5). InTech: Rijeka.
53. Corrales, M., Toepfl, S., Butz, P., Knorr, D., & Tauscher, B., (2008). Extraction of anthocyanins from grape by-products assisted by ultrasonics, high hydrostatic pressure or pulsed electric fields: A comparison. *Innovative Food Science & Emerging Technologies, 9*(1), 85–91.
54. Dörnenburg, H., & Knorr, D., (1993). Cellular permeabilization of cultured plant tissues by high electric field pulses or ultra high pressure for the recovery of secondary metabolites. *Food Biotechnology, 7*(1), 35–48.
55. Tewari, S., Sehrawat, R., Nema, P. K., & Kaur, B. P., (2017). Preservation effect of high pressure processing on ascorbic acid of fruits and vegetables: A review. *Journal of Food Biochemistry, 41*(1), e12319–n/a.
56. Bi, H. M., Zhang, S. Q., Liu, C. J., & Wang, C. Z., (2009). High hydrostatic pressure extraction of salidroside from rhodiola sachalinensis. *Journal of Food Process Engineering, 32*(1), 53–63.
57. Xiang, Y., Zhang, H., Fan, C. Q., & Yue, J. M., (2004). Novel diterpenoids and diterpenoid glycosides from siegesbeckia orientalis. *Journal of Natural Products, 67*(9), 1517–1521.
58. Sunwoo, H. H., Gujral, N., Huebl, A. C., & Kim, C. T., (2014). Application of high hydrostatic pressure and enzymatic hydrolysis for the extraction of ginsenosides from fresh ginseng root (*Panax ginseng* C.A. Myer). *Food and Bioprocess Technology, 7*(5), 1246–1254.
59. Banerjee, S., & Chatterjee, J., (2015). Efficient extraction strategies of tea (*Camellia sinensis*) biomolecules. *Journal of Food Science and Technology, 52*(6), 3158–3168.
60. Jun, X., (2009). Caffeine extraction from green tea leaves assisted by high pressure processing. *Journal of Food Engineering, 94*(1), 105–109.
61. Hawthorne, S. B., Yang, Y., & Miller, D. J., (1994). Extraction of organic pollutants from environmental solids with sub- and supercritical water. *Analytical Chemistry, 66*(18), 2912–2920.
62. Šalplachta, J., & Hohnová, B., (2017). Pressurized hot water extraction of proteins from Sambucus nigra L. branches. *Industrial Crops and Products, 108*, 312–315.
63. Vergara-Salinas, J. R., Vergara, M., Altamirano, C., Gonzalez, Á., & Pérez-Correa, J. R., (2015). Characterization of pressurized hot water extracts of grape pomace: Chemical and biological antioxidant activity. *Food Chemistry, 171*, 62–69.
64. Wu, C., Xu, H., Héritier, J., & Andlauer, W., (2012). Determination of catechins and flavonol glycosides in Chinese tea varieties. *Food Chemistry, 132*(1), 144–149.
65. Susanti, E., Ciptati, R. R., & Aulannium, R. A., (2015). Qualitative analysis of catechins from green tea GMB-4 clone using HPLC and LC-MS/MS. *Asian Pacific Journal of Tropical Biomedicine, 5*(12), 1046–1050.
66. Van der Hooft, J. J. J., Akermi, M., Ünlü, F. Y., Mihaleva, V., Roldan, V. G., Bino, R. J., et al., (2012). Structural annotation and elucidation of conjugated phenolic compounds in black, green, and white tea extracts. *Journal of Agricultural and Food Chemistry, 60*(36), 8841–8850.

67. Bastos, D. H., Saldanha, L. A., Catharino, R. R., Sawaya, A., Cunha, I. B., Carvalho, P. O., et al., (2007). Phenolic antioxidants identified by ESI-MS from Yerba Maté (Ilex paraguariensis) and green tea (Camelia sinensis) extracts. Molecules, 12(3), 423–432.
68. Wang, Y., Wei, X., & Jin, Z., (2009). Structure analysis of an acidic polysaccharide isolated from green tea. Natural Product Research, 23(7), 678–687.
69. Wang, Y., Wei, X., & Jin, Z., (2009). Structure analysis of a neutral polysaccharide isolated from green tea. Food Research International, 42(5), 739–745.
70. Scoparo, C. T., de Souza, L. M., Dartora, N., Sassaki, G. L., Gorin, P. A. J., & Iacomini, M., (2012). Analysis of Camellia sinensis green and black teas via ultra high-performance liquid chromatography assisted by liquid–liquid partition and two-dimensional liquid chromatography (size exclusion × reversed phase). Journal of Chromatography A, 1222, 29–37.
71. Scoparo, C. T., Borato, D. G., Souza, L. M., Dartora, N., Silva, L. M., Maria-Ferreira, D., et al., (2014). Gastroprotective bio-guiding fractionation of hydro-alcoholic extracts from green- and black-teas (Camellia sinensis). Food Research International, 64, 577–586.
72. Engelhardt, U. H., (2010). Chemistry of Tea. In: Comprehensive Natural Products II (pp. 999–1032). Elsevier: Oxford.
73. Chen, Z. Y., Zhu, Q. Y., Wong, Y. F., Zhang, Z., & Chung, H. Y., (1998). Stabilizing effect of ascorbic acid on green tea catechins. Journal of Agricultural and Food Chemistry, 46(7), 2512–2516.
74. Rasheed, A., & Haider, M., (1998). Antibacterial activity of Camellia sinensis extracts against dental caries. Archives of Pharmacal Research, 21 (3), 348–352.
75. Gupta, D., & Kumar, M., (2017). Evaluation of in vitro antimicrobial potential and GC–MS analysis of Camellia sinensis and Terminalia arjuna. Biotechnology Reports, 13, 19–25.
76. Choi, J. H., Yoon, S. K., Lee, K. H., Seo, M. S., Kim, D. H., Hong, S. B., et al., (2006). Antitumor activity of cell suspension culture of green tea seed (Camellia sinensis L.). Biotechnology and Bioprocess Engineering, 11(5), 396.
77. Bingfen, X., Zongchao, L., Qichao, P., Yongju, L., Xiurong, S., Likai, W., et al., (1994). The anticancer effect and anti-DNA topoisomerase II effect of extracts of camellia ptilophylla chang and camellia sinesis. Chinese Journal of Cancer Research, 6(3), 184–190.
78. Tanwar, A., Chawla, R., Ansari, M. M., Neha, Thakur, P., Chakotiya, A. S., Goel, R., et al., (2017). In vivo anti-arthritic efficacy of Camellia sinensis (L.) in collagen induced arthritis model. Biomedicine & Pharmacotherapy, 87, 92–101.
79. Balaji, G., Chalamaiah, M., Hanumanna, P., Vamsikrishna, B., Jagadeesh Kumar, D., & Venu babu, V., (2014). Mast cell stabilizing and anti-anaphylactic activity of aqueous extract of green tea (Camellia sinensis). International Journal of Veterinary Science and Medicine, 2(1), 89–94.
80. Chattopadhyay, P., Besra, S. E., Gomes, A., Das, M., Sur, P., Mitra, S., & Vedasiromoni, J. R., (2004). Anti-inflammatory activity of tea (Camellia sinensis) root extract. Life Sciences, 74(15), 1839–1849.

81. Van der Merwe, J. D., Joubert, E., Richards, E. S., Manley, M., Snijman, P. W., Marnewick, J. L., et al., (2006). A comparative study on the antimutagenic properties of aqueous extracts of *Aspalathus linearis* (rooibos), different *Cyclopia* spp. (honeybush) and *Camellia sinensis* teas. *Mutation Research/Genetic Toxicology and Environmental Mutagenesis*, *611*(1), 42–53.
82. Wu, S. C., Yen, G. C., Wang, B. S., Chiu, C. K., Yen, W. J., Chang, L. W., et al., (2007). Antimutagenic and antimicrobial activities of pu-erh tea. *LWT – Food Science and Technology*, *40*(3), 506–512.
83. Jo, Y. H., Yuk, H. G., Lee, J. H., Kim, J. C., Kim, R., & Lee, S. C., (2012). Antioxidant, tyrosinase inhibitory, and acetylcholinesterase inhibitory activities of green tea (*Camellia sinensis* L.) seed and its pericarp. *Food Science and Biotechnology*, *21*(3), 761–768.
84. Zhang, Z., Wang, X., Li, J., Wang, G., & Mao, G., (2016). Extraction and free radical scavenging activity of polysaccharide from 'Anji Baicha' (*Camellia sinensis* (L.) O. Kuntze). *International Journal of Biological Macromolecules*, *84*, 161–165.
85. Shah, S., Gani, A., Ahmad, M., Shah, A., Gani, A., & Masoodi, F. A., (2015). In vitro antioxidant and antiproliferative activity of microwave-extracted green tea and black tea (*Camellia sinensis*): A comparative study. *Nutrafoods*, *14*(4), 207–215.
86. Wang, X., Liu, Q., Zhu, H., Wang, H., Kang, J., Shen, Z., et al., (2017). Flavanols from the *Camellia sinensis* var. assamica and their hypoglycemic and hypolipidemic activities. *Acta Pharmaceutica Sinica B*, *7*(3), 342–346.
87. Gomes, A., Vedasiromoni, J. R., Das, M., Sharma, R. M., & Ganguly, D. K., (1995). Anti-hyperglycemic effect of black tea (*Camellia sinensis*) in rat. *Journal of Ethnopharmacology*, *45*(3), 223–226.
88. Abeywickrama, K. R. W., Ratnasooriya, W. D., & Amarakoon, A. M. T., (2011). Oral hypoglycaemic, antihyperglycaemic and antidiabetic activities of Sri Lankan Broken Orange Pekoe Fannings (BOPF) grade black tea (*Camellia sinensis* L.) in rats. *Journal of Ethnopharmacology*, *135*(2), 278–286.
89. Parsaei, P., Karimi, M., Asadi, S. Y., & Rafieian-kopaei, M., (2013). Bioactive components and preventive effect of green tea (*Camellia sinensis*) extract on post-laparotomy intra-abdominal adhesion in rats. *International Journal of Surgery*, *11*(9), 811–815.
90. Damiani, E., Bacchetti, T., Padella, L., Tiano, L., & Carloni, P., (2014). Antioxidant activity of different white teas: Comparison of hot and cold tea infusions. *Journal of Food Composition and Analysis*, *33*(1), 59–66.
91. Kelebek, H., (2016). LC-DAD–ESI-MS/MS characterization of phenolic constituents in Turkish black tea: Effect of infusion time and temperature. *Food Chemistry*, *204*, 227–238.
92. Ng, H. S., Teoh, A. N., Lim, J. C. W., Tan, J. S., Wan, P. K., Yim, H. S., et al., (2017). Thermo-sensitive aqueous biphasic extraction of polyphenols from *Camellia sinensis* var. assamica leaves. *Journal of the Taiwan Institute of Chemical Engineers.79*, 151–157.
93. Martins, A., Schimidt, H. L., Garcia, A., Colletta Altermann, C. D., Santos, F. W., Carpes, F. P., et al., (2017). Supplementation with different teas from *Camellia sinen-*

sis prevents memory deficits and hippocampus oxidative stress in ischemia-reperfusion. *Neurochemistry International, 108*, 287–295.
94. Jin, L., Li, X. B., Tian, D. Q., Fang, X. P., Yu, Y. M., Zhu, H. Q., et al., (2016). Antioxidant properties and color parameters of herbal teas in China. *Industrial Crops and Products, 87*, 198–209.
95. Kanda, H., Li, P., & Makino, H., (2013). Production of decaffeinated green tea leaves using liquefied dimethyl ether. *Food and Bioproducts Processing, 91*(4), 376–380.
96. Ramalho, S. A., Nigam, N., Oliveira, G. B., de Oliveira, P. A., Silva, T. O. M., dos Santos, A. G. P., et al., (2013). Effect of infusion time on phenolic compounds and caffeine content in black tea. *Food Research International, 51*(1), 155–161.
97. Horanni, R., & Engelhardt, U. H., (2013). Determination of amino acids in white, green, black, oolong, pu-erh teas and tea products. *Journal of Food Composition and Analysis, 31*(1), 94–100.
98. Asadi, S. Y., Parsaei, P., Karimi, M., Ezzati, S., Zamiri, A., Mohammadizadeh, F., et al., (2013). Effect of green tea (*Camellia sinensis*) extract on healing process of surgical wounds in rat. *International Journal of Surgery, 11*(4), 332–337.
99. Rodrigues, M. J., Neves, V., Martins, A., Rauter, A. P., Neng, N. R., Nogueira, J. M. F., et al., (2016). In vitro antioxidant and anti-inflammatory properties of *Limonium algarvense* flowers' infusions and decoctions: A comparison with green tea (*Camellia sinensis*). *Food Chemistry, 200*, 322–329.
100. Song, Y., & Yoo, S. H., (2017). Quality improvement of a rice-substituted fried noodle by utilizing the protein-polyphenol interaction between a pea protein isolate and green tea (*Camellia sinensis*) extract. *Food Chemistry, 235*, 181–187.
101. Ghoreishi, S. M., & Heidari, E., (2013). Extraction of Epigallocatechin-3-gallate from green tea via supercritical fluid technology: Neural network modeling and response surface optimization. *The Journal of Supercritical Fluids, 74*, 128–136.
102. Villanueva, B. D., Ibáñez, E., Reglero, G., Turner, C., Fornari, T., & Rodriguez-Meizoso, I., (2015). High catechins/Low caffeine powder from green tea leaves by pressurized liquid extraction and supercritical antisolvent precipitation. *Separation and Purification Technology, 148*, 49–56.
103. Vuong, Q. V., Tan, S. P., Stathopoulos, C. E., & Roach, P. D., (2012). Improved extraction of green tea components from teabags using the microwave oven. *Journal of Food Composition and Analysis, 27*(1), 95–101.
104. Bermejo, D. V., Ibáñez, E., Reglero, G., & Fornari, T., (2016). Effect of cosolvents (ethyl lactate, ethyl acetate and ethanol) on the supercritical CO_2 extraction of caffeine from green tea. *The Journal of Supercritical Fluids, 107*, 507–512.
105. Park, H. S., Im, N. G., & Kim, K. H., (2012). Extraction behaviors of caffeine and chlorophylls in supercritical decaffeination of green tea leaves. *LWT – Food Science and Technology, 45*(1), 73–78.
106. Bermejo, D. V., Mendiola, J. A., Ibáñez, E., Reglero, G., & Fornari, T., (2015). Pressurized liquid extraction of caffeine and catechins from green tea leaves using ethyl lactate, water, and ethyl lactate+water mixtures. *Food and Bioproducts Processing, 96*, 106–112.

107. Gadkari, P. V., & Balaraman, M., (2015). Extraction of catechins from decaffeinated green tea for development of nanoemulsion using palm oil and sunflower oil based lipid carrier systems. *Journal of Food Engineering, 147,* 14–23.
108. Bhushani, J. A., Kurrey, N. K., & Anandharamakrishnan, C., (2017). Nanoencapsulation of green tea catechins by electrospraying technique and its effect on controlled release and in-vitro permeability. *Journal of Food Engineering, 199,* 82–92.
109. Cui, Y., Yang, X., Lu, X., Chen, J., & Zhao, Y., (2014). Protective effects of polyphenols-enriched extract from Huangshan Maofeng green tea against CCl4-induced liver injury in mice. *Chemico-Biological Interactions, 220,* 75–83.
110. Song, R., Kelman, D., Johns, K. L., & Wright, A. D., (2012). Correlation between leaf age, shade levels, and characteristic beneficial natural constituents of tea (*Camellia sinensis*) grown in Hawaii. *Food Chemistry, 133*(3), 707–714.
111. Wulandari, R. A., Haraguchi, N., Nakayama, H., Furukawa, Y., Tanaka, T., Kuonu, I., Kawaruma, D., & Ishimaru, K., (2011). HPLC and HPLC-TOFMS analyses of tea catechins and teadenols. *Journal of Food Chemistry Safety, 18*(2), 116–121.
112. Xiao-Qin, S., Gao-Ju, Z., Yan, M., Mao, C., Sheng-Hu, C., Shuang-Mei, D., Jin-Qiong, W., et al., (2016). Isolation, identification, and biotransformation of teadenol a from solid-state fermentation of Pu-erh tea and *in vitro* antioxidant activity. *Applied Sciences, 6*(161), 1–12.
113. Ozmihci, S., (2017). Performance of batch solid-state fermentation for hydrogen production using ground wheat residue. *International Journal of Hydrogen Energy, 42,* 23494–23499.
114. Soccol, C. R., Ferreira da Costa, E. S., Junior Letti, L. A., Karp, S. G., Woiciechowski, A. L., & Porto de Souza Vandenberghe, L., (2017). Recent developments and innovations in solid state fermentation. *Biotechnology Research and Innovation, 1(1),* 52–71.
115. Alahmad, A. H., Al Fathi, H., & Alkhalaf, W., (2018). Study the influence of culture conditions on rennin production by *Rhizomucor miehei* using solid-state fermentations. *Journal of Genetic Engineering and Biotechnology, 16(1),* 213–216.

CHAPTER 9

POLYEMBRYONY IN PLANTS AND ITS POTENTIAL IN THE FOOD INDUSTRY

IXTACCIHUATL CYNTHIA GONTES-PÉREZ,[1]
JOSÉ ESPINOZA-VELÁZQUEZ,[2]
GUILLERMO CRISTIAN GUADALUPE MARTÍNEZ-ÁVILA,[3]
CRISTÓBAL N. AGUILAR[1] and RAÚL RODRÍGUEZ-HERRERA[1]

[1] *Autonomous University of Coahuila, Food Research Department, Chemistry Faculty, Blvd. V. Carranza and Jose Cardenas S/N, Republic East, ZIP 25280, Saltillo, Coahuila, Mexico, E-mail: raul.rodriguez@uadec.edu.mx*

[2] *Agrarian Autonomous University Antonio Narro, Mexican Maize Institute, Calzada Antonio Narro 1923, Buenavista, ZIP 25315, Saltillo, Coahuila, Mexico*

[3] *Autonomous University of Nuevo Leon, Agronomy School, Francisco I. Madero S/N, Hacienda el Canada, ZIP 66050, General Escobedo, Nuevo Leon, Mexico*

ABSTRACT

Polyembryony is a characteristic that is expressed in many plants. Despite that this trait has been studied further since the last century, there is still a lack of information about its development and origins. Nowadays, there is an established classification of the different types of polyembryony, as well as the causes of its development in different plant species. Additionally, it has been proved that this phenomenon can be advantageous in production, exploitation, or application of the plants in which this trait is expressed. Due to the particular characteristics of polyembryony, in this chapter there will be presented several evidences of its occurrence among some angiosperms and gymnosperms in general, and specifically the occurrence in some plants with potential on the food industry.

9.1 INTRODUCTION

Polyembryony is defined as the production of multiple embryos from a single seed [1, 2]. This characteristic has been reported as a phenomenon expressed in several organisms, both plants and animals. The presence of polyembryony in different species has been discovered in the last decades, although research about this trait is still very low in comparison to other important traits. It was in 1719, when polyembryony was first reported by Anton van Leeuwenhoek, when he found seeds in *Citrus* which contained two embryos [2]. Since that discovery, scientists have been working with more species that express this trait in nature, in order to understand this phenomenon, which can offer potential benefits for better commercial exploitation of this trait in different plant species. Based on these antecedents, it is important to understand the possible causes of polyembryony, as well as the classification of this phenomenon. Besides, in this document, there will be shown evidences of studied cases of the apparition of polyembryony in diverse plants with potential utilization in the food industry.

9.2 TYPES OF POLYEMBRYONY

Since the information of the causes of polyembryony were limited in the past two centuries; it was until the study performed by Braun [3] that it was established the first classification system, which explained four possible ways of the formation of adventives embryos. Later on, some other classifications were reported [2, 4, 5], which were based on the number of embryo sacs in which occurred the development of the adventive embryos. However, these classifications would differ in terms of the definition of true and false polyembryony.

True and false polyembryony would differ according to the development of embryos. The former consists of two or more embryos emerging from the same embryo sac, originated either from the zygote or embryo, the synergid, the antipodal cell or the nucellus or integument. In the latter, the development of embryos occurs in more than one embryo sac in the same ovule or placenta [6]. Subsequently, more authors established various classification systems based on Ernst's adding some modifications. As a result, Lakshmanan and Ambegaokar [7] proposed one system represented below in Figure 9.1.

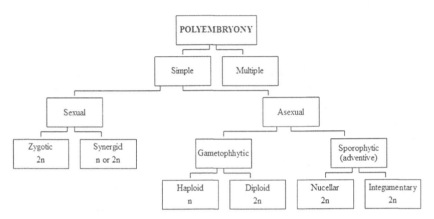

FIGURE 9.1 Classification proposed by Lakshmanan and Ambegaokar [7].

In the most general sense, polyembryony can be divided in two groups: simple and multiple.

9.2.1 SIMPLE POLYEMBRYONY

Simple polyembryony can be sexual or asexual. Sexual polyembryony can be expressed when the embryos originate from the egg cell and synergid. Otherwise, asexual embryos develop within the embryo sac without fertilization. When embryos grow from the nucellar and integumentary cells without the interpolation of the gametophytic phase, the polyembryony is called "adventive" or "sporophytic" [7].

9.2.2 MULTIPLE POLYEMBRYONY

Multiple polyembryony refers that accessory embryos are produced from two or more embryo sacs in the same ovule. Every embryo that develops in each embryo sac coincides with the scheme explained for simple polyembryony. Besides, multiple gametophytes may grow in an ovule if, during ontogeny, two or more ovules fuse and function as a single ovule. Moreover, in multiple embryo sacs within an ovule, although the egg cell (diploid) of the unreduced embryo sac may develop parthenogenetically, the egg cell (haploid) of reduced embryo sac may develop after fertilization forming a diploid-diploid twin [7].

The classification published by Peter [6] divides polyembryony in two broad groups: Spontaneous and induced, which are described as the ones that happen naturally and experimentally, respectively. In this system, it is suggested that spontaneous polyembryony is divided into two categories [8]: (1) gametophytic, originating from any gametic cell of the embryo sac after or without fertilization; and (2) sporophytic: Originating from the zygote, proembryo or the initial sporophytic cells of the ovule, namely nucellus or integuments. Despite there are more accepted systems for polyembryony classification, it is necessary to keep in mind the different characteristics that every plant may present, such as origin of the initial cells, development of the embryo, and the genetic characteristics to have the maximum understanding of this phenomenon.

9.3 CAUSES OF POLYEMBRYONY IN DIFFERENT PLANT SPECIES

In the last century, the study of polyembryony in different species has given information about origins and development of this phenomenon. As it was mentioned before, Leeuwenhoek first reported the apparition of polyembryony [2]; nonetheless, the study and discovery of the causes of polyembryony developed more than a century later. First, no clear explanation was established to describe the causes of polyembryony; however, with the help of several studies (morphological, physiological, and genetical), different possible causes of polyembryony have been proposed.

The most actualized information remarks that polyembryony, at least in plants, may happen because of several factors including those directly related to the plant such as the genetics of pollinator, amount, viability and availability of pollen, and plant nutrition. For its part, environmental factors such as air temperature, soil humidity, and wind speed also affects apparition of polyembryony [9]. In contrast, other theories about polyembryony are presented in Table 9.1; nevertheless, the causes may still be debatable to this day.

TABLE 9.1 Polyembryony Causes Adapted From the Reports by Yildiz et al. [9]

Polyembryony causes	Reference
Hormonal imbalance.	[68]
Cytological, disturbance in chromosome balance between the embryo, endosperm and mother tissue may lead to degeneration of the embryo.	[69]
	[70]
Genetic causes.	[10]
	[71]
Hybridization.	[4]
Hereditary factors.	[72]
Plant factors such as age, nutrition status, among others.	[73]

Diverse plants species have expressed polyembryony at one point. Some of them have an important development in the food industry, and can be used because of the different benefits that develop alongside with apparition of polyembryony. Seed plants, also known as Spermatophytes, are classified in two classes: Angiospermae (plants with ovules completely enclosed) and Gymnospermae (plants with naked seeds) [11]. It has been reported in the past century the occurrence of polyembryony in both groups, which involve many examples of several plants which are presented below.

9.3.1 ANGIOSPERMS

The phenomenon of polyembryony has been studied in more detail in the last century; it is reported that this characteristic is more frequent in gymnosperm plants than angiosperms [10]. Flowering plants show the unique phenomenon of double fertilization where the presence of polyembryonic angiosperms are a feature according to reports of Rangaswamy [12], where some families exhibited the characteristic. One of the most frequent types of polyembryony reported for angiosperms is the nucellar polyembryony, which can take place before or after pollination and fertilization of the egg cell [7]. Reports from Rangaswamy [12] indicate that 16 species of crassinucellate families of angiosperms develop this peculiarity.

In the Orchidaceae family, one of its species, *Zuexine sulcata* develops nucellar polyembryony since they have a low fertility pollen rate; the two or three plant adventive embryos are generated from the nucellar epidermis [13]. In contrast, *Spiranthes cernua* develops the embryos in sterile ovules but only after fertilization of fertile ovules of the plant [14]. On the other hand, *Mammilaria tenuis* from the Cactaceae family, has nucellar polyembryony as a regular event where a few nucellar embryos develop after fertilization of the flower, however, only one reaches maturity [15]. Another type of polyembryony is the zygotic polyembryony, in this case, the embryo development results from the cells produced by the zygote. Reports indicate that some species from the *Liliaceae* family, such as *Tulipa gesneriana* [16] and *Erythronium dens-canis* [17] develop this particularly type of polyembryony. Newer evidences of the apparition of polyembryony in angiosperms include *Posidonia oceanica* [18], different *Handroanthus* species such as *Handroanthus chrysotrichus* and *H. ochraceus* [19, 20], *Hevea brasiliensis* [21], some species of *Anemopaegma* [22], ornamental plants such as *Rudbeckia bicolor* [23], and some species of *Allium* genus [24].

9.3.2 GYMNOSPERMS

Polyembryony has been widely reported for gymnosperms since 1935, when Chamberlain reported numerous cases of polyembryony, as well as a system of classification that included cleavage polyembryony. In this type of polyembryony two groups may be distinguished, determinate and indeterminate. The former refers to one embryo which is considered in more favorable conditions and is usually the one that full develops; while in the latter, there are no advantages for any embryos [25]. Cleavage polyembryony occurs most frequently in gymnosperms plants [26]. Cleavage polyembryony occurs after the creation of the first series of embryonal tubes, when the lower end of the embryo system cleaves into four distinct vertical series of cells, each representing an independently developing embryo [27]. For this type of polyembryony, Teryokhin [28] established four groups of polyembryony, being the simple (occurs when several eggs of one gametophyte are independently fertilized to develop separate embryos) and cleavage (cell strands in the secondary zygote stratify to form several embryos) the two that are the most widespread. Some species of conifers that exhibit cleavage polyembryony are *Pinus, Cedrus, Tsuga, Keteleeria,* and *Abies* [29, 30]. Furthermore, simple polyembryony is common to species of the genera such as *Picea, Larix, Pseudolarix,* and *Pseudotsuga* [30, 31].

Newer evidences of the apparition of polyembryony in angiosperms include more species from the *Pinus* genus, such as *Pinus massoniana* [32], and *Pinus sibirica* [33].

9.4 POLYEMBRONY IN PLANT SPECIES WITH POTENTIAL IN THE FOOD INDUSTRY

Polyembryony has been found in different plant species around the world; although some of these reports are from the last century, there are new data of presence of polyembryony in species that must be taken into account since they can be utilized as a result of its exceptional characteristics. Each one of these plant species has characteristics those are advantageous in the production and exploitation of plant cultivars. The following evidences have been chosen to describe several examples of plant species, which were taken into consideration because of its potential in the agronomic field and its importance in the food industry in general.

9.4.1 CITRUS

Polyembryony is a common trait in *Citrus*. Since it was first discovered in 1719, it has been focus of study for the different species that appear in this group. More recent works have been directed to genetic characterization, as well as studies in non-apomictic genotypes [34]. Also, polyembryony and nucellar apomixis are important traits for citrus rootstocks breeding and selection, and for commercial multiplication. From the genetically point of view, different studies have focused in the characterization of genomic sequence in diverse *Citrus* species. That is the case of the study performed by Nakano et al. [35] where the genomic characterization showed strong relation with polyembryony in several species of this group, which can provide a basis to comprehend the mechanisms of apomixis and embryogenesis in *Citrus*. Besides, Nakano et al. [36] complemented the work with characterization of genes associated with polyembryony and *in vitro* somatic embryogenesis, which may lead to methods that facilitate the production of hybrids. Following this line of study, stress-related genes have been associated to nucellar polyembryony in *Citrus sinensis* [37], which again complements the work by Nakano et al. [35] in relation to the genes controlling polyembryony. In words of these authors, this work may be "helpful to identify genes located outside the polyembryony locus whose co-expression might be necessary to induce nucellar embryony in other species."

As it was mentioned before, polyembryony has a commercial importance in *Citrus* genus since all commercial citrus rootstocks are polyembryonic and propagated by seeds. Hence, studies have also intended to understand how zygotic or tetraploid rootstock could affect a citrus rootstock selection [38]. Another example of taking advantage of polyembryony is the case of *Citrus limon*. This species is considered slightly or moderately polyembryonic (25 to 43% of seeds being polyembryonic), for that reason, it was important to study a method to rescue the embryos at an immature phase, all of this to exploit it at its maximum. Pérez-Tornero and Porras [39] tested two different mediums and combinations of them for growth of the embryos. Finally, they concluded that this condition affected the embryo survival, germination percentage, and radical development. Moreover, polyembryony has been reported in species such as *C. jambhiri* and *Sikkim mandarin*, considered highly polyembryonic. These species can produce significantly high number of multiple seedlings which may support a better propagation of genetically uniform healthy plants [40].

9.4.2 MAIZE

Although there are few reports about polyembryony in maize, it is a known phenomenon in this specific crop. The most recent studies move back at least 10 years, and they include the characterization of plantlets, study of germplasm, besides, heredity, genetic, morphologic, and even anatomic analyses. These features are related to polyembryony and may provide a nutritional potential or production improvement for one of the most consumed food commodities in the world. Musito et al. [41] studied the germination, polyembryony frequency and seedlings abnormality;

showing acceptable germination for all the groups analyzed, with high polyembrionic frequency, and a relation between the crosses. Polyembryony heredity was also studied by Rebolloza et al. [42], where the hypothesis proposed and supported by the work states that "polyembryony occurring in two maize population is under the control of two epistatic loci, where a dominant allele at either gene pair produces the normal seedling trait (duplicate gene action), so that polyembryony is shown only by the double homozygous recessive genotype."

On the other hand, nutritional quality was approached when two different germplasm sources (a high polyembryony population used as a source of protein plus high oil content population) were combined in order to achieve a high oil content and protein quality. One of the doses analyzed (50:50) showed higher nutritional quality than normal maize, where polyembryony trait contributed to higher lysine values [43]. Besides, a study of anatomy and morphology of seedlings derived of polyembryony caryopsis was made by Espinosa-Velázquez et al. in 2012, the authors concluded that polyembryony trait influences the development of the seminal root system forcing the growth of two to four radicles.

9.4.3 MANGO

In recent years, it has been studied the polyembryonic expression in mangoes, as well as the approach to obtain some commercial benefits from the expression of this phenomenon. Cordeiro et al. [44] worked on identification of the genetic origin of polyembryony in mango (*Mangifera indica* L.) seeds. When commercialized, mango is propagated with an appropriated method to conserve the genetic characteristics, which is why it is important to use a polyembryonic variety since they produce a zygotic and nucellar plantlets. However, it is not clear the origin of each plantlet, which is why RAPD markers were used to identify which plantlets come from each type of polyembryony. Following this approach, the study published by Rocha et al. [45] similarly search for an identification of the genetic origin of seedlings of polyembrionic seeds. Besides, it is aimed to relate this characteristic with the vigor of each plantlet. As a result, it was found that the zygotic polyembryony was related with the most vigorous seedlings, despite of all the genetic diversity. Furthermore, it has been reported that since polyembryony is common in mangoes, it must be used to ease rapid multiplication and exploit its singular characteristics [46].

9.4.4 VARIOUS

Among other species that have shown polyembryony, there are several which are being studied lately. The occurrence of polyembryony in olive cultivars was first reported by Trapero et al. [47], in which spontaneous sexual polyembryony was described and characterized for cultivated olive (*Olea europaea* L.). This study can be useful for future genetic and breeding studies, and for understanding of genetic, epigenetics and environmental factors impact on polyembryony. Another example is polyembrionic rice (*Oryza sativa* L.) ApIII. Mu et al. [48] performed a morphological, cytological and anatomical investigation on the early development of embryo and endosperm in polyembryonic rice. Moreover, polyembryony has also been reported in wheat [49]; however, more studies must be made to fully comprehend the phenomena in this cereal.

9.5 ADVANTAGES OF POLYEMBRYONY IN DIVERSE CROPS

As it was mentioned before, presence of polyembryony can represent an advantage in different aspects; because of this, it is important to highlight the main enhancements that this trait may produce in the crops where it appears. Moreover, it is also of great importance to emphasize the research regarding polyembryony performed in the last years to understand the advancement and direction of the future works related to this subject.

9.5.1 NUTRITIONAL QUALITY IMPROVEMENT

The presence of more than one embryo in the seeds plays an important role in the nutritional quality, because it will affect the chemical composition of the seeds in order to feed the extra embryos. The relation of the nutritional quality with the polyembryony has been studied in maize, mostly because of its agronomical and economical importance around the world. This species is one of the most produced crop worldwide and one of the most consumed [50]. Valdez-Lara studied in 2005 the nutrimental content in the polyembryonic maize kernel, results showed that the content of linoleic and oleic acid, as well as the amino acids lysine and tryptophan, significantly increased in the polyembryonic kernels.

9.5.2 GENETIC IMPROVEMENT

In polyembryonic plants, identification in the early stages of the nucellar and zygotic embryos opens the possibility for genetic improvement, choosing the resulting hybrid from a cross in genetic improvement programs is among the advantages of polyembryony. Once the nucellar embryos are identified, rootstock that can maintain the homogeneity of the crop can be produced, old clones that have lost the vigor because of the constant propagation can be rejuvenated, and plants free from virus, viroids, phytoplasma, and bacteria can be obtained [51].

9.5.2.1 CROPS HOMOGENEITY

The advantages of having crop uniformity is that it permits crop scheduling, as well as efficient mechanical harvest because of the homogeneity in plants structure, and it may helps maximizing yield [52]. Based on Dos Santos [53] report, as polyembryony expression increase, also the possibilities of obtaining nucellar plants will increase, making it possible to have a uniform rootstock. There are reports of nucellar plants that have been obtained working with mango cultivars which showed polyembryony in more than 80% of their seeds [54], which can reflect uniformity in cultivars from this species.

Maize is another example where uniformity of cultivars has an essential role to achieve genetic homogeneity and genetic stability, and according to Živanović et al. [55] "genetic homogeneity refers to the presence of identical genotypes, whereas genetic stability refers to phenotypic uniformity (homeostasis) in different environments." Again, the uniformity of the crop represents an important advantage.

9.5.3 THE PRESENT OF POLYEMBRYONY

Although the research associated with polyembryony is low in comparison with many predominant topics in the food area, studies made in the last four years have widened the knowledge about this subject. A short timeline of relevant publications linked to polyembryony is presented below, being among them some works that include food-related species (Figure 9.2).

9.6 CONCLUSIONS AND PERSPECTIVES

Polyembryony has been reported in numerous plant species along the world and, in recent years, there are several reports of plants that may have different commercial potential uses, and exploitation of this trait can be considered advantageous for the food industry. It is important to understand the concept, in addition to the causes, origins and characteristics that make polyembryony a noteworthy trait. Although this chapter focused on some of the species believed more significant in the food industry in general, there may be other plants that may have a key role in different countries. One example is the case of the conifers, as it was previously mentioned that polyembryony is widely present in gymnosperms. Taking

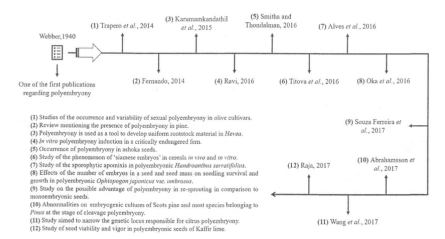

FIGURE 9.2 Timeline of recent works regarding polyembryony.

this into account, it is possible to utilize the remarkable characteristic that is useful according to each singular circumstance. For example, if polyembryony elevates the germination rates for specific species that is endangered, it can be used as a potential support for this kind of problem. For that reason, it is essential to explore in the future the different features for every plant, not only for the food industry, but for the advantages and benefits that a phenomenon such as polyembryony may produce.

ACKNOWLEDGMENTS

ICGP thanks the National Council of Science and Technology of Mexico (CONACyT) for the financial support provided during her postgraduate studies under the scholarship agreement number 582376. Financial support was received from CONACyT through the project "Identification and sequencing of DNA regions, which are controlling polyembryony in maize." FON.SEC. SEP-CONACYT CIENCIA BASICA CV–2015–03SORD2416.

CONFLICT OF INTEREST

The authors have no conflicts of interest to disclose.

KEYWORDS

- cereals
- citrus
- cultivars
- food industry
- golden apple
- maize
- mango
- olive
- plants
- polyembryony
- rice
- wheat

REFERENCES

1. Hardy, I. C. W., (1995). Protagonists of polyembryony. *Trends in Ecology & Evolution, 10*, 2.
2. Webber, J. M., (1940). Polyembryony. *The Botanical Review,* VI(31), 575–598.
3. Braun, A., (1859). About polyembryony and germination of Coelebogyne. An addendum to the treatise on parthenogenesis in plants.. *Abh. Kon. Akad. Wiss.*, 109–263.
4. Ernst, A., (1918). Bastardization as the cause of apogamy in the plant kingdom. One hypothesis to experimental heredity and evolutionary theory. *Jena G. Fischer, 70.*
5. Maheshwari, P., (1952). Polyembryony in Angiosperms. *Palaeobotanist, 1*, 319–329.
6. Peter, K. V. E., (2009). *Basics of Horticulture.* New India Publishing Agency.
7. Lakshmanan, K. K., & Ambegaokar, K. B., (1984). *Polyembryony.* In: Johri, B. M., (ed.), *Embryology of Angiosperms* (pp. 445–474). Springer: Berlin. http://doi.org/10.1007/978-3-642-69302-1_9
8. Yakovlev, M. S., (1967). Polyembryony in higher plants and principles of its classification. *Phytomorphology., 17*, 278–282.
9. Yildiz, E., Kaplankiran, M., Hakan, T., Uzun, A., & Toplu, C., (2013). Identification of zygotic and nucellar individuals produced from several citrus crosses using SSRs markers. *Not Bot Horti Agrobo., 41*(2), 478–484.
10. Maheshwari, P., & Sachar, R. C., (1963). *Polyembryony.* In: Maheshwari, P., (ed.), *Recent Advances in the Embryology of Angiosperms* (pp. 265–296). Int. Soc. Plant Morphology: Delhi.
11. Jamieson, B. G. M., & Reynolds, J. F., (1967). Tropical Plant Types. *Tropical Plant Types.* http://doi.org/10.1016/B978-0-08-012119-2.50012-8
12. Rangaswamy, N. S., (1982). Nucellus as an experimental system in basic and applied tissue culture research. *Tissue Culture of Economically Important Plants: Proceedings of the International Symposium Held at the Botany Department,* National University of Singapore, Singapore, 28–30 April 1981/Edited by A.N. Rao. COSTED and ANBS (Asian Network for Biological Sciences).
13. Swamy, B. G. L., (1946). The Embryology of *Zeuxine sulcata* Lindl. *New Phytologist., 45*(1), 132–136. http://doi.org/10.1111/j.1469-8137.1946.tb05050.x
14. Swamy, B. G. L., (1948). Agamospermy in *Spiranthes cernua. Lloydia., 11*(3), 149–162.
15. Tiagi, Y. D., (1970). Cactaceae. In: Symposium on "Comparative embryology of angiosperms." *Bull Indian Natl Sci Acad., 41*, 29–35.
16. Ernst, A., (1901). Contribution to the development of the embryo sac and the embryo (polyembryony) of *Tulipa gesneriana* L. *Flora., 88*, 37–77.
17. Guérin, P., (1930). The development of eggs and polyembryony in *Erythronium denscanis* L. *CR Acad Sci Paris., 191*, 1369–1372.
18. Balestri, E., Luccarini, G., & Lardicci, C., (2008). Abnormal embryo development in the seagrass *Posidonia oceanica. Aquatic Botany,* 89, 71–75. http://doi.org/10.1016/j.aquabot.2008.02.006
19. Bittencourt, N. S., & Moraes, C. I. G., (2010). Self-fertility and polyembryony in South American yellow trumpet trees (*Handroanthus chrysotrichus* and *H. ochra-*

ceus, Bignoniaceae): A histological study of postpollination events. *Plant Systematics and Evolution, 288*(1), 59–76. http://doi.org/10.1007/s00606-010-0313-2

20. Mendes-Rodrigues, C., Samapio, D. S., Costa, M. E., De Souza Caetano, A. P., Ranal, M. A., et al., (2012). Polyembryony increases embryo and seedling mortality but also enhances seed individual survival in *Handroanthus* species (Bignoniaceae). *Flora, 207*(4), 264–274. https://doi.org/10.1016/j.flora.2011.10.008

21. Karumamkandathil, R., Uthup, T. K., Sankaran, S., Unnikrishnan, D., Saha, T., & Nair, S. S., (2015). Genetic and epigenetic uniformity of polyembryony derived multiple seedlings of *Hevea brasiliensis*. *Protoplasma., 252*(3), 783–796. http://doi.org/10.1007/s00709-014-0713-1

22. Firetti-Leggieri, F., Lohmann, L. G., Alcantara, S., Costa, I. R., & Semir, J., (2013). Polyploidy and polyembryony in Anemopaegma (Bignonieae, Bignoniaceae). *Plant Reproduction, 26*(1), 43–53. http://doi.org/10.1007/s00497-012-0206-3

23. Musiał, K., Koscinska-Pajak, M., Sliwinska, E., & Joachimiak, A. J., (2012). Developmental events in ovules of the ornamental plant *Rudbeckia bicolor* Nutt. *Flora., 207*, 3–9. http://doi.org/10.1016/j.flora.2011.07.015

24. Specht, C. E., Meister, A., Keller, E. R. J., Korzun, L., & Börner, A., (2001). Polyembryony in species of the *Allium* genus. *Euphytica., 121*, 37–44. http://doi.org/10.1023/A:1012013121656

25. Chamberlain, C. J., (1935). The Gymnosperms. *The Botanical Review, 1*(6), 183–209.

26. Batygina, T. B., & Vinogradova, G. Y., (2007). Phenomenon of polyembryony. Genetic heterogeneity of seeds. *Russian Journal of Developmental Biology, 38*(3), 126–151. http://doi.org/10.1134/S1062360407030022

27. Attree, S. M., & Fowke, L. C., (1993). Embryogeny of gymnosperms: Advances in synthetic seed technology of conifers. *Plant Cell, Tissue and Organ Culture, 35*, 1–35.

28. Teryokhin, E. S., (1991). Problems of evolution of ontogeny in seed plants. *Tr. Botan., 2*, 1–67.

29. Buchholz, J. T., (1931). The Pine embryo and the embryos of related genera. *State Acad. Sci., 23*, 117–125.

30. Chowdhury, C. R., (1962). The embryogeny of conifers: A review. *Phytomorphology, 12*, 313–338.

31. Buchholz, J. T., (1939). The embryogeny of *Sequoia sempervirens* with a comparison of the sequoias. *Amer. J. Bot., 26*, 248–257.

32. Zhen, Y., Zhao, Z., Zheng, R., & Shi, J., (2012). Proteomic analysis of early seed development in *Pinus massoniana* L. *Plant Physiology et Biochemistry, 54*, 97–104. http://doi.org/10.1016/j.plaphy.2012.02.009

33. Krutovskii, K. V., & Politov, D. V., (1995). Allozyme evidence for polyzygotic polyembryony in Siberian Stone pine (*Pinus sibirica* Dutour). *Theoretical & Applied Genetics., 90*(6), 811–818.

34. Aleza, P., Juárez, J., Ollitrault, P., & Navarro, L., (2010). Polyembryony in non-apomictic citrus genotypes. *Annals of Botany, 106*, 533–545. http://doi.org/10.1093/aob/mcq148

35. Nakano, M., Shimada, T., Endo, T., Fujii, H., Nesumi, H., Kita, M., et al., (2012). Characterization of genomic sequence showing strong association with polyembryony among diverse *Citrus* species and cultivars, and its synteny with *Vitis* and *Populus*.

Plant Science: An International Journal of Experimental Plant Biology, 183, 131–42. http://doi.org/10.1016/j.plantsci.2011.08.002.
36. Nakano, M., Kigoshi, K., Shimizu, T., Endo, T., Shimada, T., Fujii, H., & Omura, M., (2013). Characterization of genes associated with polyembryony and in vitro somatic embryogenesis in Citrus. *Tree Genetics and Genomes, 9*(3), 795–803. http://doi.org/10.1007/s11295-013-0598-8
37. Kumar, V., Malik, S. K., Pal, D., Srinivasan, R., & Bhat, S. R., (2014). Comparative transcriptome analysis of ovules reveals stress-related genes associated with nucellar polyembryony in citrus. *Tree Genetics and Genomes, 10*(3), 449–464. http://doi.org/10.1007/s11295-013-0690-0
38. Hussain, S., Curk, F., Ollitrault, P., & Morillon, R., (2011). Facultative apomixis and chromosome doubling are sources of heterogeneity in citrus rootstock trials: Impact on clementine production and breeding selection. *Scientia Horticulturae, 130,* 815–819. http://doi.org/10.1016/j.scienta.2011.09.009
39. Pérez-Tornero, O., & Porras, I., (2008). Assessment of polyembryony in lemon: Rescue and in vitro culture of immature embryos. *Plant Cell, Tissue and Organ Culture, 93*(2), 173–180. http://doi.org/10.1007/s11240-008-9358-0.
40. Kishore, K., Monika, N., Rinchen, D., Lepcha, B., & Pandey, B., (2012). Polyembryony and seedling emergence traits in apomictic citrus. *Scientia Horticulturae, 138,* 101–107. http://doi.org/10.1016/j.scienta.2012.01.035.
41. Musito Ramírez, N., Espinoza Velázquez, J., González Vázquez, V. M., Gallegos Solórzano, José, E., & De León Castillo, H., (2008). Seedling traits in maize families derived from a polyembrinic population. *Revista Fitotecnia Mexicana, 31*(4), 399–402.
42. Rebolloza Hernández, H., Espinoza Velázquez, J., Sámano Garduño, D., & Zamora Villa, V. M., (2011). Polyembryony inheritance in two experimental maize populations. *Revista Fitotecnia Mexicana, 34*(1).
43. González-Vázquez, V. M., Espinosa-Velázquez, J., Mendoza-Villareal, R., De León-Castillo, H., & Torres-Tapia, M. A., (2011). Characterization of maize germplasm that combines a high oil content and polyembryony. *Universidad Y Ciencia., 27,* 157–167.
44. Cordeiro, M. C. R., Pinto, A. C., de, Q., Campos, V. H. V., Faleiros, F. G., & Fraga, L. M. S., (2006). Identification of the genetic origin of polyembryonic seedlings of mango. *Brazilian Fruit Growing Magazine., 28*(3), 454–457.
45. Rocha, A., Fernandes Salomão, T. M., Lopes de Siqueira, D., Cruz, C. D., & Chamhum, S. L. C., (2014). Identification of " Ubá " mango tree zygotic and nucellar seedlings. *Ceres., 61*(5), 597–604. http://doi.org/http://dx.doi.org/10.1590/0034-737X201461040001
46. Krishna, H., & Singh, S. K., (2007). Biotechnological advances in mango (Mangifera indica L.) and their future implication in crop improvement — A review. *Biotechnology Advances, 25*(3), 223–243. http://doi.org/10.1016/j.biotechadv.2007.01.001
47. Trapero, C., Barranco, D., Martín, A., & Díez, C. M., (2014). Occurrence and variability of sexual polyembryony in olive cultivars. *Scientia Horticulturae, 177,* 43–46. http://doi.org/10.1016/j.scienta.2014.07.015
48. Mu, X., Jin, B., & Teng, N., (2010). Studies on the early development of zygotic and synergid embryo and endosperm in polyembryonic rice ApIII. *Flora: Morphology,*

Distribution, Functional Ecology of Plants, 205(6), 404–410. http://doi.org/10.1016/j.flora.2009.12.023
49. Titova, G. E., Seldimirova, O. A., Kruglova, N. N., Galin, I. R., & Batygina, T. B., (2016). Phenomenon of "siamese embryos" in cereals *in vivo* and *in vitro*: Cleavage polyembryony and fasciations. *Russian Journal of Developmental Biology, 47*(3), 122–137. http://doi.org/10.1134/S1062360416030061
50. Food and Agriculture Organization of the United Nations. *FAO Statistical Pocketbook 2015*. 2015.
51. Villegas, M. A., & Andrade, M. R., (2008). *Polyembryony, adventages and disadventages for vegetative propagation*. In: Cruz-Castillo, J. G., &Torres-Lima, P. A., (eds.), *Technological Approaches in Fruit Growing* (pp. 149–167). Autonomous University of Chapingo: México.
52. Janick, J., (1999). Exploitation of heterosis: Uniformity and stability. In: *The Genetics and Exploitation of Heterosis in Crops* (pp. 319–333). CIMMYT: Mexico.
53. Dos Santos, W., Da Cunha, A., Sampaio, O., & Barreto, E., (2003). "Maravilha": Uma nova seleção de tangerina "Sunki." *Rev. Bras. Frutic., 25*(2), 268–271.
54. Martínez-Ochoa, E., Del, C., Andrade-Rodríguez, M., Rocandio Rodríguez, M., & Villegas-Monter, A., (2012). Identification of zygotic and nucellar seedlings in polyembryonic mango cultivars. *Pesq. Agropec. Bras., 47*(11), 1629–1636.
55. Živanović, T., Vračarević, M., Krstanović, S., & Šurlan-Momirović, G., (2004). Selection on uniformity and yield stability in maize. *Journal of Agricultural Sciences, 49*(1), 117–130.
56. Abrahamsson, M., Valladares, S., Merino, I., Larsson, E., & Von Arnold, S., (2017). Degeneration pattern in somatic embryos of *Pinus sylvestris* L. *In Vitro Cellular and Developmental Biology – Plant., 53*(2), 86–96. http://doi.org/10.1007/s11627-016-9797-y
57. Alves, M. F., Duarte, M. O., Bittencourt, N. S., Oliveira, P. E., & Sampaio, D. S., (2016). Sporophytic apomixis in polyembryonic *Handroanthus serratifolius* (Vahl) S.O. Grose (Bignoniaceae) characterizes the species as an agamic polyploid complex. *Plant Systematics and Evolution., 302*(6), 651–659. http://doi.org/10.1007/s00606-016-1291-9
58. Espinosa-Velázquez, J., Valdés-Reyna, J., & Alcalá-Rodríguez, J. M., (2012). Morfología y anatomía de radículas múltiples en plántulas de maíz derivadas de cariopsis con poliembrionía. *Polibotánica., 33*, 207–221.
59. Fernando, D. D., (2014). The pine reproductive process in temperate and tropical regions. *New Forests., 45*(3), 333–352. http://doi.org/10.1007/s11056-013-9403-7
60. Mora-Mata, E., (2011). Calidad física, fisiológica y bioquímica en genotipos de maíz que combina poliembrionía y alto contenido de aceite. M.Sc. Thesis, Universidad Autónoma Agraria Antonio Narro, Mexico.
61. Oka, C., Itagaki, T., & Sakai, S., (2016). Effects of the number of embryos in a seed and seed mass on seedling survival and growth in polyembryonic *Ophiopogon japonicus* var. umbrosus (Asparagaceae). *Botany., 94*(4), 261–268. http://doi.org/10.1139/cjb-2015-0214

62. Raja, K., (2017). Polymorphism influences seed viability and vigour in polyembryonic Kaffir lime (*Citrus hystrix*) seeds. *Seed Science and Technology, 45*(1), 189–197. http://doi.org/https://doi.org/10.15258/sst.2017.45.1.15
63. Ravi, B. X., (2016). In vitro polyembryony induction in a critically endangered fern, *Pteris tripartita* Sw. *Asian Pacific Journal of Reproduction, 5*(4), 345–350. http://doi.org/10.1016/j.apjr.2016.06.012
64. Smitha, G. R., & Thondaiman, V., (2016). Reproductive biology and breeding system of *Saraca asoca* (Roxb.) De Wilde: A vulnerable medicinal plant. *SpringerPlus., 5*(1), 2025. http://doi.org/10.1186/s40064-016-3709-9
65. Souza Ferreira, D. N., Camargo, J. L. C., & Ferraz, I. D. K., (2017). Multiple shoots of *Carapa surinamensis* seeds: Characterization and consequences in light of post-germination manipulation by rodents. *South African Journal of Botany, 108*, 346–351. http://doi.org/10.1016/j.sajb.2016.08.015
66. Valdez-Lara, E. L., (2005). *Ganancia en calidad nutrimental del grano como respuesta asociada a la selección para poliembrionía en maíz*. M.Sc. Dissertation, Universidad Autónoma Agraria Antonio Narro, México.
67. Wang, X., Xu, Y., Zhang, S., Cao, L., Huang, Y., Cheng, J., et al., (2017). Genomic analyses of primitive, wild and cultivated citrus provide insights into asexual reproduction. *Nature Genetics, 49*(5), 765–772. http://doi.org/10.1038/ng.3839.
68. Leroy, J. F. (1947). Polyembryony in Citrus its ineret in culture and improvement. *Rev. Intern. Bot. Appl. 27*, 483–495.
69. Dempsey, E., & Rhoades, M. M. (*1961*). Evidence for the chiasma theory of metaphase pairing. *Maize Genet. Coop. News Letter 35*, 65–66.
70. Ganeshaiah, K. N., Uma Shaanker, R., & Joshi, N. V., (1991). Evolution of polyembryony: Consequences to the fitness of mother and offspring. *Journal of Genetics 70*, 103–127.
71. Castle, L. A., & Meinke, D. W., (1993). Embryo-defective mutants as tools to study essential functions and regulatory processes in plant embryo development. *Seminars in Cell and Developmental Biology 4*, 31–39.
72. Atabekova, A .J., (1957). Polyembryony, supernumerary cotyledons and fasciation in leguminous plants. *Bulletin Principal Botany Garden Moscow 28*, 65–70.
73. Furusato, K., Ohta, Y., & Ishibashi, K., (1957). Studies on polyembryony in citrus. *Seiken Ziho 8*, 40–48.

CHAPTER 10

NATURAL POLYMERS FROM FOOD INDUSTRIAL WASTE AS RAW MATERIAL FOR NANOSTRUCTURE PRODUCTION

ARIEL GARCÍA CRUZ,[1] RODOLFO RAMOS-GONZÁLEZ,[2]
MÓNICA LIZETH CHÁVEZ-GONZÁLEZ,[1]
JUAN A. ASCACIO VALDÉS,[1] ARTURO I. MARTÍNEZ,[3]
CRISTÓBAL N. AGUILAR,[1] JOSÉ L. MARTÍNEZ,[1]
ELDA P. SEGURA-CENICEROS,[1] and ANNA ILINÁ[1]

[1] *Autonomous University of Coahuila, 25280, Saltillo, Coahuila, México Saltillo, COAH, Mexico, E-mail: anna_ilina@hotmail.com*

[2] *CONACYT – Autonomous University of Coahuila, 25280, Saltillo, Coahuila, Mexico Saltillo, COAH, Mexico*

[3] *CINVESTAV-Saltillo, Coahuila, Mexico*

ABSTRACT

Nanostructures are one of the most important inventions because they have a lot of applications in many medical and industrial fields. Nanostructures can be used to improve the safety and quality of foods as well as to produce novel food ingredients or supplements. Biopolymers from food industry waste are useful as raw materials for nanostructures synthesis because they are abundant and non-toxic, and they give to nanostructures some specific and benefit characteristics. Different biopolymers are extracted from natural sources such as wall cell plants, crustacean, and seaweeds. This review gives important evidence about the use of cellulose, lignin, chitosan, alginate, and carrageenan to obtain nanostructures.

10.1 INTRODUCTION

Natural polymers from food industry wastes can be used as a source of chemicals with a high added value. Some of them, cellulose, lignin, chitosan, alginate, and carrageenan have industrial and medicinal applications. They are biocompatible, biodegradable, non-toxic, and abundant [1].

Cellulose is the most abundant natural polymer presents in the plant's cell walls which along with hemicellulose and lignin forming a matrix that can be destroyed to liberate sugars useful to ethanol production [2–4] and other added value products.

On the other hand, Naseem et al. [5] cited that lignin is the second most abundant renewable natural polymer after cellulose. It is the unique available natural aromatic polymer which consists of a highly branched 3-D structure with benzoic, hydroxyl, methoxy, carbonyl, and carboxyl as functional groups. The same authors mentioned that lignin gives rigidity and strength to cell walls and makes up 15–40 wt% of the woody plant's dry matter. Singh and Dhepe [6] declared that due to the high availability of lignocellulosic biomass as food and agro-industries residues, and significant lignin content in the plant's cells walls (10–25%), it could be used as raw material in the lignin production. Because of existence in the nature of the huge amount of lignocellulosic material, cellulose and lignin have a high availability and low price. Chitosan is a cationic biodegradable and biocompatible biopolymer that consists of glucosamine and N-acetyl-glucosamine. Its chemical properties depend on the chain length, acetyl groups distribution, molecular weight and the deacetylation degree [7]. Chitosan can be obtained by the deacetylation reaction of chitin, which is a biopolymer from crustacean exoskeleton. It has been used as biocatalyst, in the food industry, biomedicine, pharmacy, and agriculture [8].

Moreover, Paques et al. [9] reported that alginate is a non-toxic, biodegradable, low in cost and readily available biopolymer. It is a mucoadhesive, biocompatible, and non-immunogenic substance. Authors have mentioned that this anionic polymer is produced by brown algae and bacteria, and consists of α-L guluronic acid (G) and β-D-mannuronic acid (M) residues, linearly linked by 1,4-glycosidic linkages.

By last, carrageenan is other of the most abundant polysaccharides in the nature. It is an anionic sulfated linear polysaccharide consists of D-galactose and 3, 6-anhydro-D-galactose. The carrageenan is extracted from certain red seaweeds of the *Rhodophyceae* class and it can be used to prepare foods and cosmetics as a gelling, stabilizing and thickening agent [10].

On the other hand, nanomaterials show great potential in various disciplines, including chemistry, physics, life sciences, medicine, engineering, food and medical fields [11]. Nanomaterials can be applied to improve food safety and quality, and to produce novel food ingredients or supplements. Soto-Chilaca et al. [12] mentioned that packaging was one of the first applications of nanotechnology in the food sector: About 400–500 nanoproducts were marketed. Authors point out that nanomaterials could be used in the manufacture of 25% of the all food-packaging materials.

Biopolymers can be used as a raw material to manufacture and stabilize nanostructures, giving them specific features. Khan et al. [13] classified biopolymers into four groups, such as Polysaccharides, protein, DNA and poly(hydroxyalkanoates). Venkatesan et al. [14] reported that alginate, carrageenan, fucoidan, ulvan, and laminarin can be converted into nanoparticles (NPs) by different methods, such as ionic gelation, emulsion, and polyelectrolyte complexation. These polysaccharide-based NPs can be used as carriers for the delivery of various active molecules such as proteins, peptides, anti-cancer drugs, and antibiotics.

The present work is mainly focused on the use of five abundant biopolymers as raw material to obtain nanosystems.

10.2 NANOSTRUCTURES

The prefix 'nano-' is referred to materials with diameters (or another size measure) less than 100 nm [15]. Nanoscale is a really small scale, for instance, a human hair is approximately 80,000 nm in wide, HIV/AIDS virus is 60 nm, a bacterial could reach up to 20 nm, while the carbon nanotube has around 1 nm diameter [16]. Gilca et al. [17] have mentioned that nanotechnology allows the use of chemical, physical and biological effects that do not appear outside the nanoscale.

Nanostructures can be used as: (1)- nanomaterials: They are defined as materials at nanoscale dimension that have been partially processed and have specific and particular properties because of their size; (2) nano-intermediates: They are products that contain incorporated nanomaterials, thus they have further functional prerogatives such as coatings, superconductors, and paints; (3) non-enabled products: They are considered as final product such as phones, computers, planes and contain nanomaterials and nano-intermediate into them; (4) nanotools: They are instruments using to produce nanostructures [18].

Shukla [19] reported that applications of nanotechnology in the food industry are not just represented in foods themselves but also in the things that "surround" foods, like food packaging, food processing, and sensory systems as well as the basic food and nutrition science research. Moreover, the author mentioned several additional potential applications of food nanotechnology such as improved delivery of micronutrients and bioactive food components, controlled release of bioactive compounds, magnetic separation of biologically active substances, product traceability and food safety intervention. Nanomaterials have several applications in water treatment technologies. In recent years, catalysis and photocatalysis processes using gold nanoparticles (Au-NPs) have received great attention due to their effectiveness in degrading and mineralizing organic compounds [20].

Magnetic nanoparticles constitute a type of nanostructures. The most commonly used magnetic nanoparticles include magnetite and maghemite because of their availability, stability, low environmental impact and high magnetic susceptibility [21]. Magnetic nanoparticles are commonly composed of magnetic elements such as metal oxides like magnetite (Fe_3O_4), maghemite (γ-Fe_2O_3), cobalt ferrite ($CoFe_2O_4$), etc. Laurent et al. [24] indicated that superparamagnetic iron oxide nanoparticles with a diameter less than 100 nm and appropriate surface chemistry could have numerous applications, such as magnetic resonance imaging (MRI) contrast enhancement, tissue repair, immunoassay, detoxification of biological fluids, hyperthermia, drug delivery, and cell separation.

Gupta and Gupta [22] reported that for these applications, particles must have unified properties of high magnetic saturation, biocompatibility, and surface functions. Also, these authors notified that the nano-surface could be modified with atomic layers of organic polymers or inorganic metals (e.g., gold), as well as with oxide (e.g., silica or alumina), suitable for further functionalization by attaching bioactive molecules. According to Palkhiwala and Bakshi [23], nanoparticles are a miracle invention of the century that have found novel applications in various fields of the medicine and industries (paint, electronic, renewable energy, food processing), information technology, and others.

10.2.1 METHODS TO SYNTHESIZE NANOPARTICLES

Palkhiwala and Bakshi [23] classified methods to obtain nanoparticles as: (1) Wet chemical reaction: colloidal chemistry, hydrothermal models, sol-gel, precipitation processes; (2) Gas phases synthesis: flame pyrolysis, electroexplosion, laser ablation, high-temperature evaporation, plasma synthesis; (3) Mechanical treatments: grinding, milling, mechanical alloying; (4) Formation in place: lithography, vacuum deposition, spray coatings. On the other hand, methods to manufacture magnetic iron oxide nanoparticles are classified such as: coprecipitation, reactions in constrained environments, hydrothermal and high temperature reactions, sol-gel reactions, polyol methods, flow injection synthesis, electrochemical methods, aerosol/vapor methods and sonolysis [24].

Table 10.1 shows the data of brief relationship between methods, type of synthesized nanoparticles, and their characteristics.

Table 10.2 summaries the content of some of the reports which describe the use of cellulose, lignin, alginate, chitosan, and carrageenan for nanostructures preparation and their possible industrial and medicine applications. The main routes of application are related to the use of nanosystems to improve the properties of colloids and materials, to ensure the directed and prolonged release of different drugs and microelements, to cover magnetic nanosystems, and to provide the suitable environment for the immobilization and repetitive use of enzymes.

10.2.2 NANOSTRUCTURES USING DIFFERENT BIOPOLYMERS

10.2.2.1 CELULLOSE

In the last years, many reports were published in relation to nanostructures with cellulose. Different applications of this biopolymer have been proposed.

Cheng et al. [45] suggested an efficient and low-cost approach to preparing spherical cellulose nanocrystals through chemical hydrolysis of lyocell fibers in an ammonium persulfate solution. Under mild reaction conditions, cellulose nanoparticles were spherical with a narrow diameter

TABLE 10.1 Methods to Manufacture Different Nanoparticles and Their Characteristics

Method	Type of synthesized nanoparticles	Characteristics	Reference
Sonochemical	Aqueous suspensions of lignin	Sonication at 20 KHz and 600 W was applied to a lignin dissolution for 60 min to obtain a homogeneous and stable nano-dispersion. Authors demonstrated that at the intensity applied, the compositional and structural changes of nanoparticles obtained are not significantly modified and they only depend on the lignin nature.	[17]
Coprecipitation	Chitosan-coated magnetic nanoparticles	High yield (97%) of nanoparticles was obtained at 50°C using a one-step coprecipitation of iron oxide and chitosan. The highest concentration of amino groups was observed with the lowest applied chitosan concentration (0.125% w/v) because of the thin layer formed.	[25]
γ-ray irradiation	Nickel selenide nanoparticles	They were produced using nickel acetate and selenium dioxide and were stabilized in aqueous solution by chitosan. The chitosan concentration and γ-ray irradiation absorbed dose had a significant influence in the synthesized nanoparticles.	[26]
Combination of ball milling and polyol use	Nickel selenide nanoparticles	The efficiency of the NiSe nanoparticles as a catalyst in the p-nitroaniline and p-nitrophenol reduction using $NaBH_4$ as reducing agent were studied. The particle size, morphology and the presence of surfactant had a decisive role in the reduction process.	[27]
Gamma-ray radiolysis	Ag nanoparticles	The statistic results established a quasi-Poisson probability distribution function of the particle diameter for the γ-irradiation-induced Ag nanoparticles prepared with PVP K30 as stabilizer.	[28]
Pulsed laser ablation	Bimetallic nanoparticles of iron and silver	These bimetallic nanoparticles have both magnetic and plasmonic characteristics in colloidal suspension. They were manufactured by using a nanosecond laser source emitting a 1064 nm and a two-step laser ablation of iron and silver in pure water.	[29]
Double-jet precipitation synthesis	CaF_2 nanoparticles	CaF_2 nanoparticles (9–180 nm) with spherical or cubic shape were synthesized. The results demonstrated that the nature of the solvent and a stabilizer additive influenced in the morphology and size of CaF2 nanoparticles and played a significant role in the particle formation.	[30]

TABLE 10.2 Nanostructures Contained Biopolymers and Their Applications

Type of synthesized nanostructure	Used Biopolymer	Industrial or medical applications	References
Cellulose no magnetic nanoparticles	Cellulose	Nanoparticles, microfibrillated of cellulose and cellulose nanocrystals enhanced the rheological properties of bentonite and its filtration of water-based drilling fluids.	[31]
Ethyl cellulose nanoparticles	Cellulose	These ethyl cellulose nanoparticles could be used as nanocolloids to the stabilization of foams and emulsions due to their properties.	[32]
Cellulose nanocrystals	Cellulose	Cellulose nanocrystals could significantly decrease the cost of enzymes and other processes in the bioethanol production industry.	[33]
Lignin nanoparticles	Lignin	Lignin based nanoparticles could find potential use in functional surface coatings, nanoglue, drug delivery, and microfluidic devices.	[34]
Lignin nanoparticles	Lignin	Nanoparticles are promising candidates for drug delivery applications, cancer therapy and its diagnosis, such as magnetic targeting and magnetic resonance imaging.	[35]
Lysozyme-chitosan	Chitosan	Lysozyme-chitosan nanoparticles improve lysozyme activity against *E. coli* and *B. subtilis* and could be used in the food industry.	[36]
Chitosan-magnetic nanoparticles	Chitosan	Chitosan-coated magnetic nanoparticles were used as a targeted delivery system for Doxorubicin, an anti-cancer agent, avoiding the side effects of conventional chemotherapy.	[37]
Chitosan-Alginate	Alginate and chitosan	These nanoparticles demonstrated in vitro antimicrobial activity against *P. acnes* and also to encapsulate benzoyl peroxide in the treatment of dermatology problems.	[38]
Iron loaded alginate nanoparticles	Alginate	These nanoparticles could be used as an efficient delivery system for Fe^{2+}.	[39]
Gold nanoparticles stabilized by alginate	Alginate	A multidrug carrier with potential application to treat cancer.	[40]

TABLE 10.2 (Continued)

Type of synthesized nanostructure	Used Biopolymer	Industrial or medical applications	References
Polypyrrole/Fe$_3$O$_4$/alginate beads	Alginate	They are used as magnetic solid-phase sorbent for the extraction and enrichment of estriol, β-estradiol and bisphenol A in water samples	[41]
Paclitaxel-loaded alginate nanoparticles	Alginate	The action of breast cancer tumor cells was inhibited using Paclitaxel-loaded alginate nanoparticles.	[42]
A k-carrageenan-polypyrrole-gold nanoparticles	Carrageenan	A DNA biosensor to classify the gender of *Arowana* fish, was prepared based on k-carrageenan-polypyrrole-gold nanoparticles.	[43]
Fe$_3$O$_4$-k-carrageenan nanoparticles	Carrageenan	*Pullulanase* immobilized onto Fe$_3$O$_4$-k-carrageenan nanoparticles could be used to continuous starch processing in the food industry	[44]

distribution and had a cellulose polymorphic crystalline structure with surface carboxyl groups.

Li et al. [31] investigated the crystalline structure, surface charge, morphology, and rheological behavior of the cellulose nanoparticles such as cellulose nanofibers and cellulose nanocrystals. The effectiveness of cellulose nanoparticles, including microfibrillated cellulose and cellulose nanocrystals, in enhancing the rheological and filtration performances of bentonite water-based drilling fluids was demonstrated. The resultant nanocrystals exhibited distinctive characteristics in comparison with microfibrillated cellulose such as smaller dimensions, more negative charges on the surface, higher stability in aqueous solution, but lower viscosity, and less evident shear thinning behavior.

An easy method of "dissolution-gelation-isolation-melt extrusion" to produce regenerated cellulose nanoparticles from cellulose hydrogel and simultaneously disperse them into a polymeric matrix was carried out. According to authors, the melt extrusion process provides a shear force to break up cellulose nanoparticles into smaller ones, resulting in systems with much more liquid-like behavior, higher gel point frequency and lower viscosity than the contrast samples [46].

Beaumont et al. [47] developed a nanostructured cellulose gel obtained from the Lyocell process which has much better energy efficiency than similar cellulose gel variants and higher reactivity than cellulose nanofibrils. This latter fact suggests that it can be used as a base for further chemical or physical functionalization and it could be used as tailored and high-performance materials. Lyocell is a kind of rayon which consists of cellulose fiber made from dissolving pulp (bleached wood pulp) using dry jet-wet spinning. The US Federal Trade Commission defines Lyocell as a fiber "composed of cellulose precipitated from an organic solution in which no substitution of the hydroxyl groups takes place and no chemical intermediates are formed." It classifies the fiber as a sub-category of rayon.

Siqueira et al. [48] manufactured cellulosic nanoparticles in the form of straight stiff rod-like whiskers from sisal fibers and modified with n-octadecyl isocyanate. This latter fact allowed dispersion of the nanoparticles and the production of nanocomposite films using a casting/evaporation technique for a wide range of polymeric matrices.

Moreover, cellulose is a huge source of glucose which can be used in fermentative processes to obtain ethanol, buthanol, or other biotechnologically compounds.

10.2.2.2 LIGNIN

Lignin is the second most abundant polymer in the earth behind cellulose. The biopolymer structure contains a large number of chemical functional groups providing stability against thermal- and photooxidations in polymeric applications. Lignin has antioxidant properties due to the presence of phenolic groups into its structure [49]. Currently, around 90% of lignin is burned as energy source. However, lignin can be converted into an added-value products [17].

Lievonen et al. [34] reported that lignin-based nanoparticles could find potential applications in functional surface coatings, such as nanoglues, drug delivery, and microfluidic devices. Authors extracted lignin by a Kraft process and dissolved it into tetrahydrofuran (THF) at various concentrations. The solution was filtered (0.45 µm) and put into a bag of dialysis (6–8 kDa), which was submerged into the deionized water periodically replaced. Nanoparticles were formed 24 hours later under slow stirred.

Lignin nanoparticles were synthesized, using hydroxymethylation of 10 g of lignin, suspended in distilled water and stirred for two hours at room temperature. Afterward, 50% sodium hydroxide and 25% ammonium hydroxide was added. The mixture was stirred 2 h and formaldehyde at 37% was added. The reaction was run at 85°C for four hours in water bath. The resulting product was recovered by precipitation at pH 2 with 1 N hydrochloric acid solution and separated by centrifugation. The solid phase was washed twice with distilled water, dried and weighed. Also, other lignin-nanoparticles were synthesized by lignin epoxidation [50].

Lignin-coated nanomagnetites were synthesized by coprecipitation of iron ions II and III with 0.5 M sodium hydroxide at 25°C under vigorous stirring until pH 10.5. The precipitate was heated to 80°C for 30 min, washed several times with distilled water and dried [51].

Gonugunta et al. [52] employed lignin and 15% KOH as raw material to make 25–150 nm nanoparticles using a freeze-drying process followed by thermal stabilization and carbonization. According to the authors, potassium hydroxide addition significantly influenced in the particle size.

Pure lignin nanoparticles, iron(III)-complexed lignin, and Fe_3O_4-infused lignin nanoparticles were synthesized. Nanoparticles with round shape and narrow size distribution reduced polydispersity and good stability at pH 7.4. Authors induced hydrogen peroxide production increasing the interaction with cells, using a dose-dependent cell uptake and improving their release profiles at pH 5.5–7.4 in a sustained manner [35].

10.2.2.3 CHITOSAN

Chitosan is considered the most important and practical derivative of chitin which can be used in treatment and purification fields. The amino and hydroxyl groups of chitosan make it a potential adsorbent for dyes, macromolecules and heavy metals [53]. Also, chitosan has been used a lot in medical applications. Wang et al. [54] used chitosan-N-acetyl-L-cysteine nanoparticles to improve the delivery system and the nasal absorption of insulin.

Nanoparticles of chitosan were used as an antimicrobial agent due to the chitosan antimicrobial activity against a wide variety of microorganisms including fungi, algae, and bacteria. Nanoparticles were elaborated through a preparation method in aqueous medium avoiding the use of organic solvents. Authors comment that chitosan nanoparticles and their use as charged biopolymers to stabilize nanoparticles are a recent field for investigation in food science [12].

It was immobilized *Aspergillus niger* lipase by covalent binding on chitosan-coated magnetic nanoparticles obtained by one-step co-precipitation. Hydroxyl and amino groups were activated using glycidol and glutaraldehyde, respectively. The immobilized enzyme kept more than 80% of its initial activity after 15 hydrolytic cycles. Some advantages of glutaraldehyde in comparison with glycidol as coupling agent were demonstrated [55].

The formation of chitosan–pullulanase soluble complexes and their application in the pullulanase immobilization onto Fe_3O_4–k-carrageenan nanoparticles were studied. This fact could be related to changes in secondary structures of pullulanase and chitosan bonds. Authors indicated that the complexation behavior was mainly dependent on chitosan molecular weight. In the article was presented a way to produce immobilized pullulanase upon the chitosan–pullulanase complexation, improving the stability of pullulanase. This enzyme exhibit potential for applications in the food industry for continuous productions of syrup [7].

Using the non-toxic properties of chitosan was manufactured an effective mucoadhesive chitosan-coated alginate nanoparticles delivery system for daptomycin permeation across ocular epithelia for the treatment of bacterial endophthalmitis. Authors demonstrated that daptomycin-loaded chitosan-coated alginate nanoparticles had a suitable size for ocular applications and until 92% of encapsulation efficiency [56].

Superparamagnetic spheroidal nanoparticles (10–11 nm) coated with chitosan were prepared. The synthesis was performed in one step by the coprecipitation method and in the presence of different chitosan concentrations. Nanoparticles could removal up to 53.6% of Pb^{2+} from a $PbCl_2$ aqueous dissolution. The ratio chitosan/magnetic nanoparticles were estimated between 135.9–584.8 mg/g [57].

Trichoderma reesei cellulases were immobilized covalently on 8 nm chitosan-coated magnetic nanoparticles using glutaraldehyde as a coupling agent. The immobilized cellulase retained about 80% of its activity after 15 cycles of carboxymethylcellulose hydrolysis and was easily separated with the application of an external magnetic field, which allows the reuse of the enzyme for the hydrolysis of lignocellulosic biomass [58]. This latter fact could be used to reduce the cost of the enzymatic hydrolysis process of the pretreatment lignocellulosic material to produce ethanol 2G.

Nanoparticles and chitosan nanocapsules incorporated with lime essential oil were synthesized. They showed higher antibacterial activity than nanocapsules against *Shigella dysenteriae*. Nanocapsules presents higher average size and Z-potential than nanoparticles. On the contrary, nanoparticles had a higher antibacterial activity against four food-borne pathogens than nanocapsules. This is the first report in the literature about the lime essential oil incorporated into chitosan nanoparticles and nanocapsules as well as their combined antibacterial activity [59].

Using the ionic gelation technique authors prepared chitosan nanoparticles (476.2–548.1 nm) with Lysozyme integrated achieving an increase of the antibacterial activity against *E. coli* and *B. subtilis*. This fact could be explained due to nanoparticles has a positive charge and are small, providing to both the chitosan nanoparticles and the lysozyme-chitosan nanoparticles possibility to penetrate the cell membrane breaking the cytoplasm and conducting to cell death. The authors suggested using lysozyme-chitosan nanoparticles by direct addition or incorporation into packaging in the food industry [36].

Kaempferol loaded chitosan nanoparticles as possible quorum sensing inhibitor was investigated. Authors obtained a loading and encapsulation efficiency of kaempferol into chitosan nanoparticles up to 78 and 93%, respectively. During 30 storage days, these synthesized nanoparticles inhibit the production of violacein pigment in *C. violaceum* CV026 [60].

10.2.2.4 ALGINATE

Alginate is extracted from brown algae through treatment with aqueous alkali solutions, typically with sodium hydroxide. The extract is filtered and calcium chloride is mixed with the filtrate to precipitate alginate. Then, alginate salt can be converted into alginic acid by treatment with dilute hydrochloric acid. The water-soluble sodium alginate is produced after purification [1].

Paques et al. [9] cited that alginate nanoparticles formation is based on two methods: Complex formation which can occur in an aqueous solution forming alginate nano-aggregates or on the interface of an oil droplet, forming alginate nanocapsules. Authors also mentioned that complexation can also occur through mixing alginate with an oppositely charged polyelectrolyte such as an alginate-in-oil emulsification formed by chitosan and alginate nanoparticles.

Superparamagnetic iron nanoparticles stabilized with alginate were synthesized by a modified coprecipitation method. Authors mentioned that superparamagnetic alginate nanoparticles were stable in size 12 months at 4°C and had 40 emu/g, which represents the 73% of the total solid [61].

Alginate–Fe^{2+}/Fe^{3+} polymer coated Fe_3O_4 magnetic nanoparticles with a core-shell structure were synthesized. The iron-loaded alginate nanoparticles (15–30 nm) were obtained by mean of a controlled ionic gelation method with an optimized iron encapsulation of 70% at 0.06% Fe (w/v) and 1%, w/w alginate to develop an efficient ferrous iron delivery system. According to the authors, ferrous loaded alginate nanoparticles provides an attractive delivery system for conventional oral iron therapy [39].

Also, alginate nanoparticles have medical applications such as streptomycin-loaded with poly(lactic-co-glycolic acid) (PLGA)-alginate nanoparticles using in the therapeutic treatment of salmonella infections [62].

Nanoparticles containing alginate and paclitaxel represent an innovative nanoscale delivery system that could help to the treatment of the breast cancer. As the authors explained, nanoparticles may reduce the unwanted side effects of paclitaxel (PTX), increasing drug solubility and dispersibility. Also, nanoparticles offer biocompatibility, improve conjugation with other bioactive molecules, and increase the thermal, optical, chemical and photostability of the drug [42].

Bunkoed et al. [41] prepared magnetite nanoparticles incorporated into alginate beads and coated with a polypyrrole adsorbent using an easy way of synthesis, cost-effective and good reproducibility. They could be reused 16 times as an effective magnetic solid-phase extraction sorbent for the high-efficiency extraction of chemicals from water samples and other less polar organic compounds.

Alginate is also useful for the manufacture of nanoparticles. A curcumin and methotrexate conjugated biopolymer stabilized with gold nanoparticles was synthesized in two simple reaction steps using alginate and water. This system may find potential application in targeted combination chemotherapy to treat cancer due to it showed improve cytotoxic potential against C6 glioma and MCF–7 cancer cells and it was found to be highly hemocompatible [40].

10.2.2.5 CARRAGEENAN

Carrageenan, a lineal sulphated polysaccharides extracted from red edible seaweeds, have been used as a biodegradable polymer matrix and stabilizing agent for nanoparticles. These authors prepared antimicrobial bionanocomposite films with κ-carrageenan and silver nanoparticles. The nanocomposite films exhibited characteristic antimicrobial activity against pathogenic bacteria as gram-positive (*Listeria monocytogenes*) and gram-negative (*E. coli* O157:H7) [63].

Also, pullulanase was immobilized onto Fe_3O_4-k-carrageenan nanoparticles. Authors reported that activity retention and the amount of a load of enzyme were 95.5 and 96.3% respectively, under optimal conditions. Also, the pullulanase retained 61% of the residual activity after ten consecutive reuses. According to the authors, immobilized pullulanase could be used to continuous starch processing applications in the food industry [44].

An interesting biopolymer composite film was manufactured dissolving in water (containing PEG), k-carrageenan, polyvinyl pyrrolidone, polyethylene glycol and silver nanoparticles. Dissolution was poured onto dishes and dried overnight at 40°C. Authors concluded that silver nanoparticles induced in films significant hydrophilicity, higher thermal stability, strength and lower swelling behavior. Also, this kind of film had higher fungal activity against *Aspergillus* sp., *Pencillin* sp. and *F. oxysporum* than fluconazole [64].

A magnetic and pH-sensitive nanocomposite beads were prepared modified magnetic Fe_3O_4 nanoparticles by a binary mixture of k-carrageenan and carboxymethyl chitosan. As the authors declared, the shape of magnetic nanoparticles was affected by the ratio of biopolymers. The nanoparticles swelling was affected both by the number of magnetic nanoparticles and the weight ratio of these two biopolymers. The Bovine serum albumin (BSA) adsorption kinetic study over these nanocomposites fits well to the pseudo-second-order kinetic and the Langmuir isotherm model. Authors point out that the maximum adsorption of BSA onto nanoparticles was closed to the isoelectric point of BSA (73.3 mg BSA/g adsorbent), and the adsorption process was spontaneous and exothermic [65].

Silver and magnetite nanofillers were produced in modified k-carrageenan hydrogels and it was found that nanocomposites hydrogels can improve the drug (using as a model) release in the intestine. The cross-linking and nanofiller loading favors the liberation process [66].

A biocompatibility and biodegradability polymeric carrier were prepared to mix chitosan and k-carrageenan at different pH (3–6) and mass ratios to form nanoparticles through polyelectrolyte complexation. Authors cited that the yield and the swelling percentage of the nanoparticle reached 80 and 200%, respectively. These nanostructures could be used as active ingredient delivery vehicle for prolonged release applications [67].

10.2.3 CHALLENGES TO USE BIOPOLYMERS AND NANOSYSTEMS

Humanity needs to keep a balance between environmental protection and the benefits of the industrial revolution to increase the living standards for most societies and preserve the environment for future generations. Due to humankind has caused extinctions of animals and plants, pollution of the atmosphere and environment, and other impacts, some researchers have claimed the actual geological era as the Anthropocene [68]. Application of recent nanotechnologies lacks in-depth studies of their impacts on the environment and humans.

Researchers are concern regarding the interaction of nanometric materials and biological systems [69, 70], because of nanoparticles have some negative impacts for the human body when they are used in drug delivery systems, for example, pulmonary and alveolar inflammation and the effect of the autonomic imbalance on heart and vascular function [69]. The interface between nanoparticles and cell membranes depend on different factors such as the electrostatic interaction, particle size, surface area and others [70], due to this more studies are needed.

On the other hand, the use of biopolymers can often reduce the negative impacts of nanomaterials, improving some properties of nanosystems. For instance, chitosan is a well-known biocompatible and biodegradable biopolymer, appropriate for transport into mucosal routes. It has been used in ophthalmic formulations for topical application [71]. Both chitosan and alginate have food and pharmaceutical industrial applications and they are used as nanocarriers for the release of therapeutic proteins. Das et al. synthesized an ALG-CS-PF127 composite NPs useful to formulate nanosystems of hydrophobic drugs against cancer cells. Chitosan, alginate, and ALG-CS-PF127 have biodegradable and nontoxic properties [72]. Good biocompatibility of nanoparticle systems assurances little irritation, satisfactory bioavailability and adequate compatibility with the ocular tissues, when they are used at appropriate and narrow particle size range [73]. It is important to consider nanoparticles size distribution as one of the main limitations to access a specific organ or tissue. Biocompatibility, biodegradability and the abundant existence of renewable sources are some properties that highlight the advantages of biopolymers compared to synthetic polymers when used in drug and gene delivery [74].

Polymer industry tends to sustainability using biodegradable biopolymers from renewable resources. The use of biopolymers as an alternative to artificial plastics is a promising idea mainly for practical, economic applications or to get low environment impact. However, the current nanotechnological use of biopolymers is at the beginning of the industrial development [75]. To scale up the production of biopolymeric nanoparticles much effort is necessary to do, even in laboratory studies to define their cytotoxicity, biodegradability, biocompatibility, and influence on immune response for applications in clinical drug delivery [76] and other uses related with biological systems.

10.3 CONCLUSIONS

Natural polymers from food industry wastes can be used as a source of chemicals with a high added value related to their use for nanostructures synthesis. Nanostructures could be used as a part of new industrial and medical applications. Cellulose, lignin, chitosan, alginate, and carrageenan are biopolymers using in the manufacture of the different nanostructure and they have relatively low-cost because of they are extracted from natural sources. Methods to make nanoparticles are four mainly: Wet chemical, gas phases synthesis, mechanical and formed in place. Using natural polymers to manufacture nanoparticles will improve nanotechnology systems forward to many medical and industrial applications and will reduce the cost of several processes.

ACKNOWLEDGMENTS

Authors want to acknowledgments to the Autonomous University of Coahuila, Mexico, and the CONACYT (National Council for Science and Technology of Mexico).

KEYWORDS

- cellulose
- chitosan
- lignin
- nanoparticles

REFERENCES

1. Venkatesan, J., Bhatnagar, I., Manivasagan, P., Kang, K., & Kim, S., (2014). Alginate composites for bone tissue engineering: A review. *Int. J. Biol. Macromol., 72C*, 269–281.
2. García, A., Cara, C., Moya, M., Rapado, J., Puls, J., Castro, E., & Martín, C., (2014). Dilute sulphuric acid pretreatment and enzymatic hydrolysis of jatropha curcas fruit shells for ethanol production. *Ind. Crops Prod., 53*, 148–153.

3. García, A., López, Y., Keikhosro, K., Benitez, A., Lundin, M., Mohammad, T., et al., (2015). *Chemical and Physical Characterization and Acid Hydrolysis of a Mixture of Jatropha Curcas., 49*, 737–744.
4. Martín, C., García, A., Schreiber, A., Puls, J., & Saake, B., (2015). Combination of water extraction with dilute-sulphuric acid pretreatment for enhancing the enzymatic hydrolysis of jatropha curcas shells. *Ind. Crops Prod., 64*, 233–241.
5. Naseem, A., Tabasum, S., Zia, K. M., Zuber, M., Ali, M., & Noreen, A., (2016). Lignin-derivatives based polymers, blends and composites: A review. *Int. J. Biol. Macromol., 93*, 296–313.
6. Singh, S. K., & Dhepe, P., L., (2016). Isolation of lignin by organosolv process from different varieties of rice husk: Understanding their physical and chemical properties. *Bioresour. Technol., 221*, 310–317.
7. Long, J., Xu, E., Li, X., Wu, Z., Wang, F., Xu, X., et al., (2016). Effect of chitosan molecular weight on the formation of chitosan-pullulanase soluble complexes and their application in the immobilization of pullulanase onto Fe3O4-κ-carrageenan nanoparticles. *Food Chem., 202*, 49–58.
8. Aranaz, I., Gutiérrez, M. C., Ferrer, M. L., & Del Monte, F., (2014). Preparation of chitosan nanocomposites with a macroporous structure by unidirectional freezing and subsequent freeze-drying. *Mar. Drugs, 12*(11), 5619–5642.
9. Paques, J. P., Van Der Linden, E., Van Rijn, C. J. M., & Sagis, L., M. C., (2014). Preparation methods of alginate nanoparticles. *Adv. Colloid Interface Sci., 209*, 163–171.
10. Prajapati, V. D., Maheriya, P. M., Jani, G. K., & Solanki, H., K., (2014). Carrageenan: A natural seaweed polysaccharide and its applications. *Carbohydr. Polym., 105*(1), 97–112.
11. Khan, I., & Oh, D. H., (2015). Integration of nisin into nanoparticles for application in foods. *Innov. Food Sci. Emerg. Technol., 34*, 376–384.
12. Soto-chilaca, G. A., Ramírez-corona, N., Palou, E., & López-malo, A., (2016). Food Antimicrobial agents using phenolic compounds, chitosan, and related nanoparticles. *J. Food Bioengineering and Nanoprocessing, 1*(2), 161–185.
13. Khan, A. K., Saba, A. U., Nawazish, S., Akhtar, F., Rashid, R., Mir, S., Nasir, B., Iqbal, F., Afzal, S., Pervaiz, F., & Murtaza, G., (2017). Carrageenan based bionanocomposites as drug delivery tool with special emphasis on the influence of ferromagnetic nanoparticles. *Oxidative Medicine and Cellular Longevity*, 2–13.
14. Venkatesan, J., Anil, S., Kim, S. K., & Shim, M., (2016). Seaweed polysaccharide-based nanoparticles: Preparation and applications for drug delivery. *Polymers, 8*(2), 2–25.
15. Yiu, H. H. P., & Keane, M., A., (2012). Enzyme-magnetic nanoparticle hybrids: New effective catalysts for the production of high value chemicals. *J. Chem. Technol. Biotechnol., 87*(5), 583–594.
16. Záyago-Lau, E., & Foladori, G., (2010). Nanotechnology in Mexico: An uncertain development. *Econ. Soc. y Terrti., 10*(32), 143–178.
17. Gilca, I. A., Popa, V. I., & Crestini, C., (2015). Obtaining lignin nanoparticles by sonication. *Ultrason. Sonochem., 23*, 369–375.

18. Zayago, L. E., Foladori, G., Appelbaum, R., & Arteaga Figueroa, E., R., (2013). Nanotechnology companies in Mexico: Towards a first inventory. *Estud. Soc.*, *21*(42), 9–25.
19. Shukla, K., (2012). Nanotechnology and emerging trends in dairy foods: The inside story to food additives and ingredients. *Int. J. Nano. Sci & Tech.* *1*(1), 41–58.
20. Ayati, A., Ahmadpour, A., Bamoharram, F. F., Tanhaei, B., Mänttäri, M., & Sillanpää, M., (2014). A review on catalytic applications of Au/TiO$_2$ nanoparticles in the removal of water pollutant. *Chemosphere*, *107*, 163–174.
21. Indira, T., & Lakshmi, P., (2010). Magnetic nanoparticles: A review. *Int. J. Pharm.*, *3*(3), 1035–1042.
22. Gupta, A. K., & Gupta, M., (2005). Synthesis and surface engineering of iron oxide nanoparticles for biomedical applications. *Biomaterials*, *26*(18), 3995–4021.
23. Palkhiwala, S., & Bakshi, S., (2014). Engineered nanoparticles: Revisiting safety concerns in light of ethno medicine. *AYU (An Int. Q. J. Res. Ayurveda)*, *35*(3), 237.
24. Laurent, S., Forge, D., Port, M., Roch, A., Robic, C., Vander Elst, L., & Muller, R., N., (2008). Magnetic iron oxide nanoparticles: Synthesis, stabilization, vectorization, physicochemical characterizations and biological applications. *Chem. Rev.*, *108*(6), 2064–2110.
25. Osuna, Y., Gregorio, K. M., Gaona, G., de la Garza, I., Ilyna, A., Diaz, E., et al., (2012). Chitosan-coated magnetic nanoparticles with low chitosan content prepared in one-step. *J. Nanomater.*, 1–7.
26. Huang, N. M., Radiman, S., Lim, H. N., Yeong, S. K., Khiew, P. S., Chiu, W. S., et al., (2009). ??-Ray assisted synthesis of Ni3Se2 nanoparticles stabilized by natural polymer. *Chem. Eng. J.*, *147*(2–3), 399–404.
27. Subbarao, U., Marakatti, V. S., Amshumali, M. K., Loukya, B., Singh, D. K., Datta, R., et al., (2016). Size and morphology controlled NISE nanoparticles as efficient catalyst for the reduction reactions. *J. Solid State Chem.*, *244*, 84–92.
28. Liang, J., Liang, S. S., Wen Ye, S., Ying Ye, Meng. L., (2015). Prediction of size Distribution of Ag nanoparticles synthesized via gamma-ray radiolysis. *Radiat. Phys. Chem.*, *114*, 5–11.
29. Muniz-Miranda, M., Gellini, C., Giorgetti, E., & Margheri, G., (2016). Bifunctional Fe3O4/Ag nanoparticles obtained by two-step laser ablation in pure water. *J. Colloid Interface Sci.*, *489*, 100–105.
30. Safronikhin, A., Ehrlich, H., & Lisichkin, G., (2017). Double-jet precipitation synthesis of CaF2 nanoparticles: The effect of temperature, solvent, and stabilizer on size and morphology. *J. Alloys Compd.*, *694*, 1182–1188.
31. Li, M. C., Wu, Q., Song, K., Qing, Y., & Wu, Y., (2015). Cellulose nanoparticles as modifiers for rheology and fluid loss in bentonite water-based fluids. *ACS Appl. Mater. Interfaces*, *7*(8), 5009–5016.
32. Bizmark, N., & Ioannidis, M., A., (2015). Effects of ionic strength on the colloidal stability and interfacial assembly of hydrophobic ethyl cellulose nanoparticles. *Langmuir*, *31*(34), 9282–9289.
33. Durán, N., Lemes, A. P., Durán, M., Freer, J., & Baeza, J., (2011). A minireview of cellulose nanocrystals and its potential integration as co-product in bioethanol production. *J. Chil. Chem. Soc.*, *56*(2), 672–677.

34. Lievonen, M., Valle-Delgado, J. J., Mattinen, M. L., Hult, E. L., Lintinen, K., & Kostiainen, M., L., (2015). Simple process for lignin nanoparticle preparation. *Green Chem.*, 1416–1422.
35. Figueiredo, P., Lintinen, K., Kiriazis, A., Hynninen, V., Liu, Z., Bauleth-Ramos, T., & Rahikkala, A., L., (2017). In vitro evaluation of biodegradable lignin-based nanoparticles for drug delivery and enhanced antiproliferation effect in cancer cells. *Biomaterials, 121*, 97–108.
36. Wu, T., Wu, C., Fu, S., Wang, L., Yuan, C., Chen, S., L., (2017). Integration of lysozyme into chitosan nanoparticles for improving antibacterial activity. *Carbohydr. Polym., 155*, 192–200.
37. Unsoy, G., Khodadust, R., Yalcin, S., Mutlu, P., & Gunduz, U., (2014). Synthesis of Doxorubicin loaded magnetic chitosan nanoparticles for pH responsive targeted drug delivery. *Eur. J. Pharm. Sci., 62*, 243–250.
38. Friedman, A. J., Phan, J., Schairer, D. O., Champer, J., Qin, M., & Pirouz, A., L., (2012). Antimicrobial and anti-inflammatory activity of chitosan-alginate nanoparticles: A targeted therapy for cutaneous pathogens. *J. Invest. Dermatol., 133*(10), 1231–1239.
39. Katuwavila, N. P., Perera, A. D. L. C., Dahanayake, D., Karunaratne, V., Amaratunga, G. A. J., & Karunaratne, D. N., (2016). Alginate nanoparticles protect ferrous from oxidation: Potential iron delivery system. *Int. J. Pharm., 513*(1–2), 404–409.
40. Dey, S., Sherly, M. C. D., Rekha, M. R., & Sreenivasan, K., (2016). Alginate stabilized gold nanoparticle as multidrug carrier: Evaluation of cellular interactions and hemolytic potential. *Carbohydr. Polym., 136*, 71–80.
41. Bunkoed, O., Nurerk, P., Wannapob, R., & Kanatharana, P., (2016). Polypyrrole-coated alginate/magnetite nanoparticles composite sorbent for the extraction of endocrine-disrupting compounds. *J. Sep. Sci., 39*(18), 3602–3609.
42. Markeb, A. A., El-maali, N. A., Sayed, D. M., Osama, A., Abdel-malek, M. A. Y., Zaki, A. H., et al., (2016). Preclinical efficacy of a novel paclitaxel-loaded alginate nanoparticle for breast cancer treatment. *Int. J. Breast Cancer, 2016.* 1–8.
43. Esmaeili, C., Heng, L. Y., Chiang, C. P., Rashid, Z. A., Safitri, E., & Malon, M. R. S. P., (2017). A DNA biosensor based on kappa-carrageenan-polypyrrole-gold nanoparticles composite for gender determination of arowana fish (Scleropages Formosus). *Sensors Actuators, B Chem., 242*, 616–624.
44. Jie Long, Zhengzong, Wu, Xingfei, Li, Enbo, Xu, Xueming, Xu, & Zhengyu, Jin, A. J., (2015). Magnetic (Fe3O4−κ-Carrageenan) nanoparticles by electrostatic. pdf., pp. 3534–3542.
45. Cheng, M., Qin, Z., Liu, Y., Qin, Y., Li, T., Chen, L., & Zhu, M., (2014). Efficient extraction of carboxylated spherical cellulose nanocrystals with narrow distribution through Hydrolysis of lyocell fibers by using ammonium persulfate as an oxidant. *J. Mater. Chem. A, 2*(1), 251.
46. Zhang, L. Q., Niu, B., Yang, S. G., Huang, H. D., Zhong, G. J., & Li, Z., M., (2016). Simultaneous preparation and dispersion of regenerated cellulose nanoparticles using a facile protocol of dissolution-gelation-isolation-melt extrusion. *ACS Sustain. Chem. Eng., 4*(5), 2470–2478.

47. Beaumont, M., Rennhofer, H., Opietnik, M., Lichtenegger, H. C., Potthast, A., & Rosenau, T., (2016). Nanostructured cellulose II gel consisting of spherical particles. *ACS Sustain. Chem. Eng.*, *4*(8), 4424–4432.
48. Siqueira, G., Bras, J., & Dufresne, A., (2010). New process of chemical grafting of cellulose nanoparticles with a long chain isocyanate. *Langmuir*, *26*(1), 402–411.
49. Jawaid, M., Sapuan, M., & Alotman, O., (2017). *Green Biocomposities. Design and Applications*. XII, pp 345.
50. Popa, V. I., Căpraru, A. M., Grama, S., & Măluțan, T., (2011). Nanoparticles based on modified lignins with biocide properties. *Cellul. Chem. Technol.*, *45*(3–4), 221–226.
51. Kołodyńska, D., Gęca, M., Pylypchuk, I. V., & Hubicki, Z., (2016). Development of new effective sorbents based on nanomagnetite. *Nanoscale Res. Lett.*, *11*(1), 1–13.
52. Gonugunta, P., Vivekanandhan, S., Mohanty, A. K., & Misra, M., (2012). A study on synthesis and characterization of biobased carbon nanoparticles from lignin. *World J. Nano Sci. Eng.*, *2*, 148–153.
53. Salehi, E., Daraei, P., & Arabi, S. A., (2016). A review on chitosan-based adsorptive membranes. *Carbohydrate Polymers*, 419–432.
54. Wang, X., Zheng, C., Wu, Z., Teng, D., Zhang, X., Wang, Z., & Li, C., (2009). Chitosan-NAC nanoparticles as a vehicle for nasal absorption enhancement of insulin. *J. Biomed. Mater. Res. – Part B Appl. Biomater.*, *88*(1), 150–161.
55. Osuna, Y., Sandoval, J., Saade, H., López, R. G., Martinez, J. L., Colunga, E. M., et al., (2015). Immobilization of aspergillus niger lipase on chitosan-coated magnetic nanoparticles using two covalent-binding methods. *Bioprocess Biosyst. Eng.*, *38*(8), 1437–1445.
56. Costa, J. R., Silva, N. C., Sarmento, B., & Pintado, M., (2015). Potential chitosan-coated alginate nanoparticles for ocular delivery of daptomycin. *Eur. J. Clin. Microbiol. Infect. Dis.*, *34*(6), 1255–1262.
57. Gregorio, K. M., Pineda, M. G., Rivera, J. E., Hurtado, G., Saade, H., Martínez, J. L., et al., (2012). One-step method for preparation of magnetic nanoparticles. Coated with chitosan. *Nanomaterials*, 1–8.
58. Sanchez, J., Martínez, J. L., Segura, P., López, G., Saade, H., Medina, M. A., et al., (2016). Cellulases immobilization on chitosan-coated magnetic nanoparticles : Application for agave atrovirens lignocellulosic biomass hydrolysis. *Bioprocess Biosyst. Eng.*, 1–14.
59. Sotelo-Boyás, M. E., Correa-Pacheco, Z. N., Bautista-Baños, S., & Corona-Rangel, M. L., (2017). Physicochemical characterization of chitosan nanoparticles and nanocapsules incorporated with lime essential oil and their antibacterial activity against food-borne pathogens. *LWT – Food Sci. Technol.*, *77*, 15–20.
60. Ilk, S., Salam, N., Zgen, M., & Korkusuz, F., (2017). Chitosan nanoparticles enhances the anti-quorum sensing activity of kaempferol. *Int. J. Biol. Macromol.*, *94*, 653–662.
61. Ma, H., Qi, X., Maitani, Y., & Nagai, T., (2007). Preparation and characterization of superparamagnetic iron oxide nanoparticles stabilized by alginate. *Int. J. Pharm.*, *333*(1–2), 177–186.
62. Asadi, A., (2014). Streptomycin-loaded PLGA-alginate nanoparticles: Preparation, characterization, and assessment. *Appl. Nanosci.*, *4*(4), 455–460.

63. Rhim, J. W., & Wang, L. F., (2014). Preparation and characterization of carrageenan-based nanocomposite films reinforced with clay mineral and silver nanoparticles. *Appl. Clay Sci.*, *97–98*, 174–181.
64. Fouda, M. M. G., El-Aassar, M. R., El Fawal, G. F., Hafez, E. E., Masry, S. H. D., & Abdel-Megeed, A., (2015). K-Carrageenan/poly vinyl pyrollidone/polyethylene glycol/silver nanoparticles film for biomedical application. *Int. J. Biol. Macromol.*, *74*, 179–184.
65. Mahdavinia, G. R., & Etemadi, H., (2015). Surface modification of iron oxide nanoparticles with κ-carrageenan/carboxymethyl chitosan for effective adsorption of bovine serum albumin. *Arab. J. Chem.*, 1–20.
66. Hezaveh, H., Muhamad, I., I., (2012). The effect of nanoparticles on gastrointestinal release from modified K-carrageenan *nanocomposite hydrogels.*, 138–145.
67. Han-Choi, Y., & Misran, M., (2016). Preparation and characterization of pH dependent κ-carrageenan-chitosan nanoparticle as potential slow release delivery carrier. *Iranian Polymer Journal.*, *25*(12), 1037–1046.
68. Vilaplana, F., Strömberg, E., & Karlsson, S., (2010). Environmental and resource aspects of sustainable biocomposites, *Polym. Degrad. Stab.*, *95*(11), 2147–2161.
69. Rashmi, C., Nesalin, J. A., Mani, T., T., & College, B., (2016). Nanoparticulate drug delivery system-Review, *Int. J. Res. Pharm. Chem.*, *6*(3), 491–500.
70. Hirano, A., Yoshikawa, H., Matsushita, S., Yamada, Y., & Shiraki, K., (2012). Adsorption and disruption of lipid bilayers by nanoscale protein aggregates, *Langmuir.*, *28*, 3887–3895.
71. Wang, X., Zheng, C., Wu, Z., Teng, D., Zhang, X., Wang, Z., & Li, C., (2008). Chitosan-NAC nanoparticles as a vehicle for nasal absorption enhancement of insulin. *J. Biomed. Mater. Res. – Part B Appl. Biomater.*, 150–161.
72. Das, R. K., Kasoju, N., & Bora, U., (2010). Encapsulation of curcumin in alginate-chitosan-pluronic composite nanoparticles for delivery to cancer cells, *Nanomedicine Nanotechnology, Biol. Med.*, *6*(1), 153–160.
73. Raviña, M., De, Paolicelli, P., Sanchez, A., Seijo, B., & Jose, M., (2010). Chitosan-based nanostructures : A delivery platform for ocular therapeutics. *Adv. Drug Deliv. Rev.*, *62*(1), 100–117.
74. Nitta S. K., & Numata, K., (2013). Biopolymer-based nanoparticles for drug/gene delivery and tissue engineering, *Int. J. Mol. Sci.*, *14*, 1629–1654.
75. Schwark, F., (2009). Influence factors for scenario analysis for new environmental technologies–the case for biopolymer technology, *J. Clean. Prod.*, *17*(7), 644–652.
76. Sundar, S., Kundu, J., & Kundu, S., C., (2010). Biopolymeric nanoparticles, *Sci. Technol. Adv. Mater.*, *11*, 1–13.

CHAPTER 11

ARSENIC AND HEAVY METAL CONTAMINATION IN GROUNDWATER IN SOUTH ASIA, GROUNDWATER REMEDIATION, AND THE VISION FOR THE FUTURE

SUKANCHAN PALIT[1,2]

[1] *Department of Chemical Engineering, University of Petroleum and Energy Studies, Bidholi via Premnagar (Post Office), Dehradun–248007, India*

[2] *#43, Judges Bagan, Haridevpur (PO), Kolkata–700082, India, Tel.: 0091-8958728093, E-mail: sukanchan68@gmail.com, sukanchan92@gmail.com*

ABSTRACT

Human civilization and human scientific endeavor today is in the path of scientific vision and scientific regeneration. Environmental engineering science is witnessing drastic and dramatic challenges and surpassing one visionary boundary over another. Arsenic and heavy metal poisoning of groundwater in developed and developing economies are the cause of immense global concern and science and technology has few answers to the huge scientific barriers and hurdles. South Asian countries such as Bangladesh and India are in the midst of immense environmental engineering turmoil. Global water shortage, lack of pure drinking water and the intricacies of innovations and scientific vision will all lead a long and visionary way in the true emancipation of environmental engineering science and chemical process engineering. Arsenic groundwater poisoning in South Asia is human civilization's largest environmental crisis. In this treatise, the author pointedly focuses on the immense scientific success, the vast technological profundity and the immediate needs of scientific vision and vast scientific innovation in tackling this issue. The authors lucidly depict the vast scientific research endeavor in the field of groundwater remediation of heavy metals, which includes arsenic also. This entire chapter covers the technologies used in groundwater remediation in India and Bangladesh. Separation science and chemical process technology are the immense scientific necessities of environmental engineering endeavor today. Human scientific progress in environmental engineering science needs to be envisioned and readdressed as research forays in groundwater remediation gains pivotal importance. Technological advancements, scientific achievements and vast scientific vision stand as major torchbearers in the research pursuit to global drinking water issues. This chapter will open new windows of scientific innovation and vast scientific ingenuity in the furtherance of science of groundwater remediation.

11.1 INTRODUCTION

Environmental engineering science today stands in the midst of deep scientific vision and scientific fortitude. Provision of pure drinking water is a scientific necessity in present-day human civilization. Today groundwater poisoning with arsenic and heavy metals is a disaster of science and engineering. Man's immense scientific vision, human mankind's vast scientific prowess and the world of scientific challenges will all lead a long and visionary way in the true emancipation of science and engineering today. Arsenic groundwater contamination is human mankind's largest environmental disaster. Technology and engineering science has few answers to this enigmatic scientific problem. South Asia particularly India and Bangladesh are in the threshold of this largest environmental engineering disaster. Arsenic groundwater contamination scenario is not confined in developing countries. It is also a major scientific challenge in developed nations. In this chapter, the author pointedly focuses on the scientific barriers and the innovations in technology which research and development initiatives faces in the course of scientific history and time. The vision and the challenge of engineering science and technology are immense in present-day human civilization. Provision of pure drinking water is of utmost necessity in the successful path towards greater emancipation of science and technology today. Water, energy and electricity are the essential parameters towards human scientific progress today. Science of groundwater remediation is today in the path of newer scientific vision and vast regeneration. Arsenic groundwater poisoning is an enigmatic issue in developing countries in South Asia particularly Bangladesh and India. In this chapter, the author deeply reiterates the scientific success, the vast scientific potential and the scientific ingenuity in groundwater remediation technologies. Today is the age of scientific innovation and vast scientific vision. The author vastly delineates the status of arsenic groundwater poisoning in South Asia and other developed nations and the health hazards linked with arsenic groundwater contamination. Technological motivation, scientific validation and human scientific endeavor's prowess will lead a long and visionary way in the true realization of environmental engineering science today. Technology and engineering science has practically no answers to the environmental devastation in Bangladesh and India. This chapter is a veritable eye-opener to the scientific intricacies and scientific vision behind groundwater remediation technologies.

11.2 THE AIM AND OBJECTIVE OF THIS STUDY

Human civilization and human scientific endeavor are today moving towards a newer path of scientific regeneration and deep scientific vision. Technology and engineering science are highly challenged today. Modern science today stands in the midst of vision and scientific fortitude. Water science and technology in the similar manner stands in the crucial juncture of scientific introspection and vast scientific vision. The primary aim and objective of this study is to target the immense scientific research forays in the field of arsenic groundwater remediation and the future of global water crisis. Heavy metal groundwater contamination also is a major thrust area of this well-researched study. Human mankind's immense scientific prowess, the scientific vision and the vast technological profundity will all lead a long and visionary way in the true emancipation of the science of environmental sustainability today. Human scientific progress, the scientific prowess of mankind and the immediate need of environmental protection will lead a long and visionary way in the true emancipation of environmental engineering science today. Human mankind needs to be envisioned as science and engineering surges forward. The primary aim and objective of this study is to bring to the scientific arena the vast scientific vision of groundwater remediation technologies and the immediate need of environmental engineering science. This chapter is a visionary and watershed text in the grave crisis on arsenic groundwater contamination scenario in Bangladesh and India. South Asia particularly Bangladesh and India are in the grip of an unending environmental crisis. Human scientific progress and human scientific regeneration are at a state of immense catastrophe. The author in this chapter repeatedly proclaims the global need to look into the immense calamity of arsenic groundwater poisoning in developing and developed nations throughout the world primarily Bangladesh and India. Environmental sustainability and the domain of environmental engineering science stands in the midst of deep scientific introspection and vast vision. Human scientific progress is today in a similar manner in a state of immense scientific distress. Technology and engineering science has few answers to the marauding domain of heavy metal and arsenic groundwater poisoning. Human mankind's immense scientific prowess and stature, the needs of human society and the vast vision of environmental protection will all lead a long and visionary way in the true emancipation of environmental sustainability today. Sustainable development is the cornerstone of scientific research today. Without sustainability, the world would be standing between crisis and catastrophe in every area of research pursuit. The visionary words of Dr. Gro Harlem Brundtland, former Prime Minister of Norway on the " science of sustainability" needs to be readdressed and re-envisioned with each step of human scientific progress. In this chapter, the author reiterates the scientific success, the scientific divination and the vast scientific discernment in the path towards heavy metal and arsenic groundwater contamination in various parts of the world.

11.3 GROUNDWATER POISONING IN SOUTH ASIA AND THE VISION FOR THE FUTURE

Groundwater poisoning in South Asia is a cause of immense concern as human civilization and human scientific endeavor moves forward. The status of groundwater poisoning in developing and developed countries needs to be seriously pondered upon as technology and engineering science treads a visionary path of vision and scientific fortitude. Human civilization and human scientific endeavor in environmental engineering science today stands in the midst of scientific comprehension and scientific travails. Technology and engineering science has few answers to the ever-increasing and marauding global issue of groundwater heavy metal poisoning. South Asia particularly Bangladesh and India are at the threshold of an unbelievable devastation as modern science and engineering surges forward. The state of environment in Bangladesh and India is the world's largest environmental disaster. Diverse branches of science and engineering are involved in the research and development initiatives in groundwater heavy metal remediation. Geology, biological sciences, chemical process engineering and environmental engineering science are the visionary avenues of research pursuit which needs to be envisioned and re-envisaged as environmental protection garners immense importance in the present-day human civilization. The challenge and the vision of science of groundwater remediation needs to be readdressed and re-structured as human scientific endeavor moves from one visionary paradigm over another. Human scientific genre today stands in the midst of deep vision and scientific fortitude. Groundwater poisoning in Bangladesh and India is veritably aggravating with the progress of science and technology. Global status of heavy metal and arsenic groundwater contamination is absolutely grave and replete with vast scientific failures. Human scientific progress needs to be re-envisioned with the passage of scientific history and visionary timeframe. Environmental engineering paradigm in the similar manner needs to be overhauled and restructured. Sustainable development is a veritable utopia in developing economies in South Asia. Technology and engineering science in arsenic groundwater remediation is highly challenged and needs to be globally streamlined. Scientific passion, the vast scientific excellence and the scientific progeny of environmental science and chemical process engineering are in the path of newer vision and newer paradigm. South Asia particularly Bangladesh and India are in the brink of an unending environmental disaster. Applied science such as chemistry, biological science and geology needs to surpass cross-boundary research. Multi-disciplinary research effort should usher in a new era in science and engineering of water technology. Human mankind's immense scientific prowess, the vast technological ingenuity and the scientific marvels of environmental engineering will all lead a long and visionary way in the true realization and true emancipation of groundwater remediation.

11.4 SCIENTIFIC VISION AND SCIENTIFIC DOCTRINE OF GROUNDWATER REMEDIATION

Scientific doctrine in the field of groundwater remediation today stands in the midst of deep scientific introspection and vast scientific profundity. Groundwater poisoning and groundwater remediation is a veritable bane of present-day human civilization. Technological vision, the scientific prowess of human mankind and the vast scientific regeneration will all lead a long and visionary way in the true emancipation of environmental engineering science today. Human scientific challenges and barriers are veritably immense as groundwater remediation gains immense importance in today's scientific paradigm. Today chemical process engineering and environmental engineering science are two opposite sides of the visionary coin. Novel separation processes such as membrane science are the remarkable avenues of scientific research pursuit today. Traditional and non-traditional techniques of environmental engineering science are opening up new vistas of scientific discernment and deep scientific vision. Membrane separation processes are the utmost need of environmental engineering science today. Scientific discerning and scientific fortitude are the hallmarks of human scientific research pursuit today. The vision for the future of environmental engineering science and chemical process engineering is immense and path-breaking. Human mankind stands today in the midst of scientific introspection and deep scientific thought. Technology and engineering science has few answers to the global enigmatic issue of groundwater arsenic and heavy metal poisoning. The scientific doctrine and the scientific vision of groundwater remediation needs to be revamped and restructured as science, technology and engineering moves forward. Human suffering due to groundwater poisoning is immense in developing and poor countries around the world. A consortium of scientists and engineers in diverse fields is of immediate necessity in tackling this global scientific enigma. Scientific rigor, academic discernment and the world of engineering marvels will lead a long and visionary way in the true emancipation and the effective realization of environmental protection science and environmental engineering today. Today is the world of nuclear science and space technology. Engineering marvels and technological validation are the pivotal points of scientific advancements today. In a similar manner groundwater remediation today is in the path of newer vision and newer advancements. Human scientific progress and scientific achievements are the cornerstones of research pursuit and scientific validation in present-day human civilization. Groundwater remediation should be a marvel of science as science and technology moves forward. This is an absolute need of the human society and a disastrous bane of human civilization. Geological Sciences, biological sciences along with diverse areas of engineering science needs to be combined in the research pursuit in groundwater remediation and decontamination. Technology has simply no answers to the marauding effect of arsenic groundwater poisoning and veritably human scientific endeavor needs to be re-envisioned with immediate effect. Thus, the author reiterates the vast scientific vision of groundwater remediation and drinking water purification. Technological innovations such as membrane science, nanofiltration and advanced oxidation processes are ushering in a new era in water treatment, drinking water treatment and the vast domain of wastewater treatment.

11.5 TECHNOLOGICAL INNOVATION AND THE NEEDS OF PROVISION OF PURE DRINKING WATER

Technological innovations are needed today as global water shortage issues wreck the vast scientific firmament. Human scientific challenges stand in the midst of vision and forbearance. Provision of pure drinking water is a veritable challenge to human civilization today. Technological vision and profundity, the vast scientific travails and the success of scientific research pursuit will all lead a long and visionary way in the true realization of science and engineering today. Provision of clean drinking water is an important parameter towards the progress of human civilization and human scientific progress today. Technology and engineering science has few answers towards the immense scientific barriers to groundwater poisoning in Bangladesh and India today. Today innovations in technology and engineering science are veritably linked with environmental sustainability and environmental protection. Technology today has few answers to the marauding and ever-growing concern for arsenic and heavy metal contamination of groundwater poisoning in South Asia. The needs of provision of pure drinking water remains an important and pivotal challenge to the human civilization and human scientific progress. Bangladesh and India are in the threshold of a vicious environmental crisis. The visionary words of Dr. Gro Harlem Brundtland, former Prime Minister of Norway on "sustainability and sustainable development" needs to be envisioned and readdressed as regards the world's largest environmental engineering catastrophe in arsenic drinking water poisoning. The needs of geological science and biological sciences assumes immense importance in tackling arsenic groundwater contamination along with environmental scientists and chemical engineers.

11.6 GLOBAL WATER SHORTAGE AND ENVIRONMENTAL SUSTAINABILITY

Global water shortage and the success of environmental sustainability are the two opposite sides of the visionary coin today. Human civilization and human scientific endeavor stands today in the crucial juncture of deep introspection and vast scientific vision. Environmental and energy sustainability today in the similar manner stands in the crucial juncture of technological profundity and vast scientific vision. Today there is an immediate need of sustainable development in human planet. Lack of pure drinking water, the loss of ecological biodiversity and frequent environmental catastrophes has urged the scientists and engineers to delve deep into the world of environmental engineering science and environmental sustainability. Human scientific vision needs to be revamped and overhauled as human civilization witnesses immense scientific struggles and deep travails. Heavy metal groundwater poisoning and arsenic groundwater contamination stands today in the midst of deep scientific vision and unending scientific introspection. Bangladesh and India stands at the threshold of a major environmental crisis. In this chapter the author instinctively ponders with deep insight the success of environmental engineering tools and the vast profundity of environmental sustainability in the path towards the furtherance of science and engineering. Global water crisis is an absolute bane to human civilization and human scientific research pursuit. Technology and engineering science has few answers to the ever-growing concern of groundwater heavy metal contamination.

11.7 SUSTAINABLE DEVELOPMENT AND MODERN SCIENCE

Modern science today is in the path of newer scientific regeneration and vast scientific rejuvenation. The challenge and the vision of environmental engineering science and chemical process engineering are changing the very path of human scientific research pursuit. Sustainable development whether it is environmental or energy are the utmost need of the hour. Human mankind's immense scientific prowess, man's vast scientific discernment and the needs of human society will lead a long and visionary way in the true realization of environmental sustainability today. Provision of basic human needs are the scientific imperatives of human civilization today. Modern science needs to be revamped with the passage of scientific history and time. Today sustainability and progress in engineering science are two opposite sides of the visionary coin. Social, economic, energy and environmental sustainability are the cornerstones of human scientific research pursuit in environmental engineering science today. Sustainability today stands in the midst of deep comprehension and vision. Environmental and energy sustainability are the veritable hallmarks of human scientific research pursuit today. Modern science is today faced with immense scientific barriers such as depletion of fossil fuel resources, climate change and loss of ecological biodiversity. In this chapter, the author pointedly focuses on the scientific success, the scientific vision and the deep scientific ingenuity behind groundwater remediation with the sole aim towards the furtherance of science and engineering. Modern science today is a huge colossus with a definite and purposeful vision of its own. Human mankind's immense scientific prowess, man's vast technological profundity and the world of scientific validation will all lead a long and visionary way in the true emancipation of environmental engineering tools. Today sustainable development is veritably linked with environmental engineering science, provision of water, electricity and food. Thus the need of environmental and energy sustainability.

11.8 ENERGY AND ENVIRONMENTAL SUSTAINABILITY AND THE NEEDS OF HUMAN SOCIETY

Energy and environmental sustainability are the hallmarks of scientific progress of human society. Energy crisis, energy security and the vast scientific vision behind sustainability are the global torchbearers towards a newer visionary era in energy engineering. Renewable energy today is a huge colossus with a definite vision of its own. Energy engineering and energy sustainability today stands in the midst of vast scientific vision and deep scientific comprehension. As arsenic and heavy metal groundwater contamination gains immense importance, environmental engineering science and environmental sustainability are the utmost need of the hour. The foundations of global environmental sustainability are today laid down by environmental engineering tools such as traditional and nontraditional techniques. Human civilization today stands in the midst of deep scientific introspection and unending environmental engineering catastrophes. The need of human society and modern science are environmental and energy sustainability. Today depletion of fossil fuel resources has denuded the scientific firmament and the vast scientific fabric of petroleum engineering science and petroleum exploration. Technological adroitness and deep scientific astuteness are the pillars of research pursuit in diverse areas of science and engineering today. Energy engineering, chemical process engineering and environmental engineering should be at the forefront of scientific research pursuit today. Today the needs of the human society are immense and path-breaking. In this chapter, the author deeply delineates the scientific success, the scientific landscape and the vast scientific firmament of groundwater contamination and its much-needed remediation. South Asia mainly Bangladesh and India are in the throes of an immense scientific challenge and an unexplained environmental crisis. The scientific issue is vicious and never-ending. Human scientific challenges and research and development initiatives needs to be re-organized and revamped with the passage of scientific history and time. Thus the immense and path-breaking need of environmental and energy sustainability.

11.9 NANOMATERIALS FOR ENVIRONMENTAL PROTECTION

Human society and human scientific research pursuit are today towards the path of immense scientific regeneration. Today, in the present-day human civilization, nanotechnology and environmental engineering science are the two opposite sides of the visionary coin. Nanomaterials and engineered materials are the smart materials of engineering science today.

Technological and scientific validation along with scientific profundity are the cornerstones of research pursuit today. Groundwater heavy metal contamination is a disaster to human civilization today. Environmental protection thus also stands in the midst of scientific introspection and vast and definite vision. Application of nanomaterials and environmental protection are today linked by a visionary umbilical cord. Thus human scientific progress needs to be envisioned and readdressed with the passage of scientific history and time. Technology today has few answers to the marauding issue of environmental protection as regards arsenic groundwater remediation. Thus the need of nanomaterials and engineered nanomaterials.

Aghabozorg et al. [1] discussed with cogent and lucid far-sightedness removal of pollutants from the environment using sorbents and nanocatalysts. Chemical and petroleum industries introduce many pollutants into the surrounding environment. These pollutants, such as aliphatic and aromatic compounds, disinfection by-products, dyes, volatile organic compounds, liquid nuclear wastes, exhaust emission from transportation vehicles, various classes of sulfur compounds, toxic heavy metal ions, and other chemicals enter the air, surface water, soil etc. and create a global problem [1]. Today arsenic polluted water is also treated by nanomaterials and nanocomposites. Technology of nanomaterials is vastly advancing. The authors also discussed removal of sulfur compounds from fuels, elimination of heavy metals and separation of the dangerous radionuclides from liquid nuclear wastes. Human scientific endeavor today stands in the midst of vision and deep comprehension. This chapter opens up new windows of innovation and scientific understanding in the field of nanomaterials and the vast domain of nanotechnology [1].

Azzaza et al. [2] discussed lucidly nanomaterials for heavy metal removal. Toxic metals(called heavy metals) are a natural part of the earth's crust. The authors in this chapter discusses sources of heavy metals in the environment, nanotechnology for environment remediation, the application of adsorbents which includes carbon-based nanomaterials, activated carbon, carbon nanotubes, graphenes and metal-based nanomaterials [2]. The other area discussed is that of nanosized metal oxides. Human scientific regeneration, the needs of nanotechnology to human society and the vast scientific profundity of nanomaterials application will lead a long and effective way in the true emancipation of nano-engineering and nanoscience today [2].

Technology and engineering science are today in the path of newer scientific rejuvenation. Human scientific progress in environmental engineering and nanotechnology are linked by an unsevered umbilical cord. The challenges and the vision of application of nanomaterials in environmental protection are immense and groundbreaking. Thus the immediate need of envisioning of water technology and nanotechnology. In this entire chapter, the author repeatedly pronounces the utmost need of drinking water treatment, the need of groundwater remediation and the vast vision of environmental engineering science.

11.10 SCIENTIFIC RESEARCH PURSUIT IN GROUNDWATER REMEDIATION

Groundwater contamination and its successive remediation are the enigma of science today. Arsenic groundwater poisoning of human mankind in developed and developing countries are shattering the veritable scientific firmament of environmental engineering science and environmental sustainability. Research and development initiatives in groundwater heavy metal remediation are today gaining immense importance as human civilization stands in the threshold of scientific vision and vast technological motivation. Today research endeavor in the field of water science and water technology are moving from one visionary paradigm to another. In this section, the author deeply enunciates the need of environmental sustainability and groundwater remediation for the future of human civilization. The research forays and the research and development initiatives in the field of arsenic and heavy metal groundwater contamination are today replete with vision and deep scientific fortitude. Technology and engineering science has few answers to the huge and enigmatic issue of provision of clean drinking water. Successive human generations need to be re-envisioned and revamped as regards the contribution of modern science in tackling global water issues. Water hiatus is an enigma of science and engineering today. The vast human progress and the needs of modern human society are today two opposite sides of the visionary coin. The author in this section reiterates the scope and the vision of this science of groundwater remediation.

Chakraborti et al. [3] deeply discussed with immense lucidity in a 20-year study report the status of groundwater arsenic contamination in the state of West Bengal, India. Technology and engineering science has today few answers to the ever-growing crisis of arsenic groundwater contamination in the state of West Bengal, India. Since 1988, the authors have analyzed 140–150 water samples from tube wells in all 19 districts of West Bengal for arsenic; 48.1% had arsenic above 10 µg/L (WHO guideline value), 23.8% above 50 µg/L (Indian standard) and 3.3% above 300 µg/L (concentrations predicting overt arsenical skin lesions) [3]. Based veritably on arsenic concentrations, the authors have classified West Bengal into three zones: Highly affected (9 districts mainly in eastern side of Bhagirathi river), mildly affected (5 districts in northern part) and unaffected (5 districts in western part). The estimated number of tube wells in 8 of the highly affected districts is 1.3 million, and estimated population drinking arsenic-contaminated water above 10 and 50 µg/L were 9.5 and 4.2 million, respectively. In West Bengal alone, 26 million people are potentially at risk from drinking arsenic-contaminated water (above 10 µg/L). By studying information for water from different depths from 107253 wells, the authors noted that arsenic contamination decreased with increasing depth [3]. Measured arsenic concentrations in two tube wells in Kolkata for 325 and 51 days during 2002–2005, showed 15% oscillatory movement along without any long term trend. Regional variability is veritably dependent on sub-surface geology. Science of arsenic groundwater remediation is today in a latent phase. Human civilization's immense scientific girth, the scientific grit and determination will today lead a long and visionary way in the true realization of environmental engineering science and water science and technology. The world of environmental engineering science is witnessing immense restructuring and definite vision. Arsenic groundwater contamination is a global environmental protection issue [3]. Bangladesh and India are the countries severely affected by arsenic groundwater contamination. Groundwater arsenic contamination and its severe health effects in South-East Asian countries came to prominence during the last decade. Bangladesh, India and China are the worse affected nations throughout the world. This survey is for 20 years in India and it indicates that some areas of all the states (Uttar Pradesh, Bihar, Jharkhand, West Bengal) in Ganga plain are arsenic affected and thousands are severely suffering from arsenic toxicity and millions are at risk [3]. Human scientific progress and human scientific ingenuity in environmental engineering science are today in a state of immense newer innovation and vast scientific rejuvenation. Environmental protection is the utmost need of the hour as science and engineering surges ahead. Arsenic groundwater remediation is the need of the moment as science struggles and visionary scientific endeavor moves towards a newer environmental engineering paradigm [3].

Guha Mazumder [4] delineates with deep and cogent insight chronic arsenic toxicity and human health. Chronic arsenic toxicity (arsenicosis) due to drinking of arsenic-contaminated groundwater is a major health hazard in developing and developed economies throughout the world including India and Bangladesh [4]. A lot of new information is emerging from extensive research on health effects of chronic arsenic toxicity in humans in the last two decades. Science and technology are huge colossus with a definite and purposeful vision of its own. Available literature has been reviewed to highlight the problem including its severe malignancies [4]. This chapter reviews the immense health hazards involved in chronic arsenic toxicity. Human suffering in chronic arsenic toxicity are untold and has few scientific answers. This chapter is replete with vast scientific discernment and scientific understanding. Human mankind's immense scientific prowess, the vast world of scientific validation and the visionary world of environmental engineering science will all lead a long and effective way in the true realization of groundwater remediation today. In this chapter, available literature has been deeply reviewed to highlight the problems including malignancies. The vast vision of science, the challenges of modern science and the needs of environmental engineering science are the torchbearers towards a newer visionary era in the field of groundwater remediation today. Pigmentation and keratosis are the skin lesions characteristics of chronic arsenic toxicity. Cancer of skin, lung and urinary bladder are important cancers associated with chronic arsenic toxicity. Stoppage of drinking of arsenic-contaminated water is the mainstay in the overall management of arsenicosis. Technology and engineering science are vast domains of scientific understanding and scientific profundity today. The author in this treatise deeply enumerates the immense scientific success and scientific ingenuity in the domain of health hazards due to arsenic groundwater remediation [4].

Dey et al. [5] deeply discussed with immense lucidity groundwater arsenic contamination in West Bengal, India, its current scenario, effects and possible ways of mitigation. During the past two decades, Arsenic contamination via groundwater has become a serious and burning issue worldwide and is now a major concern in the Indo-Bangladesh Gangetic delta [5]. Arsenic groundwater contamination is today a veritable bane to human civilization and human scientific progress. The vast world of scientific validation and the needs of environmental engineering science are the veritable pallbearers towards a newer visionary era in environmental protection today. Arsenic enters human body through contaminated groundwater and has disastrous effect to human health [5]. Food safety in this region is also facing severe consequences as bio-accumulation of Arsenic is occurring in food crops irrigated with Arsenic-contaminated groundwater. Chronic and long exposure to Arsenic can cause not only cancerous and non-cancer health effects. Human scientific endeavor and human scientific forays are in a state of immense catastrophe as arsenic groundwater poisoning devastates the veritable environmental engineering firmament. Scientific reports suggest that about 20% of the population in West Bengal, India is highly affected [5]. Various techniques and innovations are today being introduced to provide arsenic-free drinking water at an affordable cost. Scientific vision and scientific innovations are ever-growing but it has few answers to the marauding, disastrous and catastrophic issue of arsenic groundwater poisoning. It is now recognized that millions of people from India have been vastly endangered by the prospect of consuming water contaminated with arsenic at levels greater than the guideline value of acceptable level set by World Health Organization; more than 95% of them live in West Bengal. This is the world's largest environmental engineering disaster [5]. Adverse health effects of arsenic depend strongly on the dose and the duration of arsenic and other heavy metals. Chronic intake of drinking water with elevated arsenic concentrations can result in development of arsenicosis, the collective term for diseases caused by chronic exposure to arsenic and other heavy metals. It includes several kinds of skin lesions, and cancers, such as hyperpigmentation, keratosis, gangrene, cancer of different internal organs [5]. This chapter is a veritable eye-opener to the vast health hazards and the scientific profundity in tackling arsenic groundwater remediation.

Smith et al. [6] deeply with lucid and cogent insight delineated contamination of drinking-water by arsenic in Bangladesh and its public health emergency. Bangladesh is today faced with environmental disasters with immense proportions. Arsenic groundwater contamination is one such burning and enigmatic issue. The contamination of groundwater by Arsenic in Bangladesh is the largest poisoning of a population in history, with millions of people affected [6]. Science has limited and few answers to the ever-growing concern of arsenic and heavy metal groundwater poisoning. This chapter describes the history of the discovery of arsenic in drinking-water in Bangladesh and highly recommends intervention strategies. Human civilization and human scientific vision is today in a state of immense distress and deep catastrophe. The basic intervention is the identification and provision of pure drinking water. Community education and participation are essential to ensure that intervention strategies are essential. The authors in this chapter discussed with cogent insight the scientific needs, the scientific strategies and the vast scientific excellence in confronting arsenic groundwater poisoning issue. Bangladesh is today grappling with largest mass poisoning of a population in human history because groundwater used for drinking has been vastly contaminated with naturally occurring arsenic [6].

Hashim et al. [7] discussed in minute details remediation technologies for heavy metal contaminated groundwater. The challenge and the vision of remediation science are immense and evergrowing. Environmental sustainability is the hallmark of this chapter. The contamination of groundwater by heavy metal, originating either from natural soil sources or from anthropogenic sources is a matter of grave concern to the public health. Remediation of contaminated groundwater is of highest priority since billions of people around the world use it for drinking water purpose. Scientific vision, deep scientific ingenuity and vast scientific regeneration are the cornerstones of this chapter. In this chapter, 35 approaches for groundwater treatment are reviewed and classified under three large categories, viz chemical, biochemical/biosorption and physico-chemical treatment processes. Scientific validation along with technological innovations are the necessities of human scientific endeavor today. In the area of groundwater remediation, science also needs to be envisioned and readdressed. In this chapters, the authors deeply discussed heavy metal in groundwater, its sources, chemical property and speciation, groundwater remediation technologies, chemical treatment technologies, in-situ treatment, reduction by using iron-based technologies, in situ soil flushing, in-situ chemical fixation, and bio-sorption of heavy metals [7]. Modern science today stands in the midst of vision and scientific fortitude. Water purification, drinking water treatment and industrial wastewater treatment are the utmost needs of human progress today. The authors in this chapter pointedly focuses in these visionary research pursuits [7].

Shannon et al. [8] with vast far-sightedness discussed science and technology for water purification in the coming decades. One of the burning issues afflicting people throughout the world is inadequate access to clean drinking water and sanitation. The water scarcity is growing at a rapid pace globally. Addressing these problems calls out for a tremendous amount of research and development initiatives [8]. In this chapter, the authors highlights some of the science and technology developed to improve the disinfection and decontamination of water at a large scale. Water purification and environmental engineering today stands in the midst of vast scientific introspection and definite vision. This chapter is a vast eye-opener to the grave concerns of environmental protection today. The authors deeply discusses the science of disinfection, decontamination, reuse and reclamation and desalination science [8]. Global research and development initiatives in desalination are in a state of immense scientific comprehension and ingenuity. In the Middle East countries water scarcity is a cause of immense concern. The work highlighted here, plus the tremendous amount of additional effort being conducted in every continent is sowing the seeds of a revolution in water purification and treatment. The enormity of the problems facing the world from the lack of adequate clean water and sanitation means that much work needs to be done to address the challenges particular to developing nations which is in the midst of a scientific quagmire. The authors in this treatise deeply comprehends the scientific success, the scientific vision and the scientific rejuvenation in water purification technologies today [8].

Human mankind and human scientific endeavor are today in the midst of deep scientific fortitude and scientific revelation. Water science and technology needs to be envisioned and streamlined with the progress of scientific history and time. Heavy metal and arsenic groundwater remediation needs to be on the topmost priority in every water research and development initiatives. Scientific articulation and scientific vision should be the hallmarks of research and development forays into water technology today. In the similar manner, nanotechnology applications in water science should be the pivotal points in the scientific vision of tomorrow. Nanomaterials applications in water science and environmental protection needs to be revamped as science and engineering moves forward. In this entire article, the author repeatedly pronounces the vast scientific success and the scientific divination in groundwater remediation, drinking water treatment and wastewater treatment. Global water crisis and environmental catastrophes are the veritable challenges of tomorrow. This chapter will open newer windows of innovation and scientific instinct in decades to come [9–11].

11.11 FUTURE RESEARCH TRENDS AND THE PROVISION OF CLEAN DRINKING WATER

Human mankind and human scientific endeavor today are in the path of newer scientific regeneration and deep scientific barriers. The shackles and fetters of scientific research pursuit are today immense and involves vast academic rigor. Scientific validation and technological far-sightedness are the pillars of engineering science today. Human mankind's immense scientific prowess and enigma, the scientific instinct and the needs of basic necessities such as water, food, shelter and electricity will all lead a long and visionary way in the true realization of sustainable development today. Provision of clean drinking water in the similar manner is a veritable imperative towards the success of science today. Future research trends should be directed towards newer innovation and a newer paradigm. Developed and developing economies are in the deep abyss of environmental disaster and vast scientific comprehension. South Asia particularly Bangladesh and India are today witnessing immense scientific challenges in water research and development initiatives. Today technology and engineering science are witnessing immense scientific turmoil and scientific vision. The world of challenges in groundwater research and development initiatives needs to be revitalized and restructured as science and engineering surges forward. Today is the world of environmental and energy sustainability. The visionary definition of sustainability by Dr. Gro Harlem Brundtland, the former Prime Minister of Norway needs to be envisioned and re-organized as science and technology of environmental protection and environmental engineering science moves forward. Human scientific progress today is in a state of immense distress and deep scientific revelation. The future research trends should be directed towards newer innovation and newer vision. Global research and development spending should be targeted towards technological innovation [9–11].

11.12 FUTURE RESEARCH THOUGHTS AND GLOBAL SUSTAINABILITY

Global vision on sustainability is immense and groundbreaking. Human scientific progress in water science and technology are today in a state of immense catastrophe and an unending crisis. Arsenic and heavy metal contamination of groundwater are challenging veritably the scientific firmament of vision and deep comprehension. Energy and environmental sustainability are today in the path of newer scientific regeneration and deep scientific rejuvenation. The vision of science, man's immense scientific prowess and grit and human mankind's vast regeneration will all lead a long and visionary way in the true emancipation of the science of sustainable development. Technology and engineering science have few answers to the burning issue of global heavy metal groundwater contamination. In this chapter, the author repeatedly stresses upon the new innovations and the newer visionary pathways in the scientific vision of arsenic and heavy metal groundwater contamination and its remediation. South Asia particularly Bangladesh and India today are in the threshold of newer scientific fortitude and deep scientific revelation. This chapter opens up newer answers and delves deep into the vast scientific success and the immense scientific potential in the innovations and technologies in heavy metal groundwater remediation. The challenges and the vision are immense and groundbreaking. Future research thoughts needs to be streamlined and envisioned as global water research and development initiatives enters into a visionary scientific arena. Human civilization and human scientific progress today stands in the midst of deep scientific introspection and vast scientific discernment. This chapter delves deep into the needs of science, the scientific prowess and the immediate vision of environmental engineering techniques. Future research trends in the field of water technology needs to be realigned and streamlined as science and engineering moves from one visionary paradigm towards another. This chapter is a veritable eye-opener to the areas of traditional and non-traditional environmental engineering techniques. Novel separation processes such as membrane science needs to be vehemently readdressed as global water crisis multiplies. The author in this chapter deeply targets the vast scientific vision of application of environmental engineering tools such as advanced oxidation processes and integrated advanced oxidation processes. The utmost need of technology and engineering science needs to be readdressed as global water shortage and impending crisis devastates the scientific firmament. This challenge and the vast vision are related in minute details in this chapter [9–11].

11.13 CONCLUSION AND SCIENTIFIC PERSPECTIVES

Scientific prowess, scientific ingenuity and vast scientific determination are the veritable challenges of human scientific research pursuit today. Technology and engineering science has few answers to the difficulties and scientific intricacies of arsenic groundwater remediation. Scientific perspectives need to be streamlined as the science of arsenic groundwater remediation techniques enters a newer visionary era and a newer paradigm. The world of environmental engineering science is in a state of immense scientific revelation. Human mankind and human scientific research pursuit are highly challenged today as regards application of sustainability to human society. Modern science is in the path of scientific regeneration. Chemical process engineering, environmental engineering science and energy engineering needs to be vastly envisioned as science surges forward. In this chapter, the author delves deep into the scientific truth and the scientific divination in groundwater remediation techniques. Human society and modern science needs to be vastly envisioned and reorganized with the passage of scientific history and time. This chapter is a veritable eye-opener to the scientific intricacies and the scientific barriers in the field of groundwater remediation, environmental engineering techniques and novel separation processes. The situation of arsenic groundwater contamination in Bangladesh and India is extremely grave and of immense concern. The challenge and vision are equally immense and groundbreaking. Environmental engineering science today stands in the midst of deep catastrophe and scientific vision. Future research trends should be directed towards greater innovation and effective groundwater remediation technologies. Today is the world of space science and nuclear technology. Yet, water science and technology should not be neglected as human civilization surges forward towards a newer visionary age. The author in this chapter reiterates the grave concern and the utmost need of environmental engineering science with the sole aim of furtherance of science and engineering.

KEYWORDS

- arsenic
- groundwater
- nanomaterials
- poison
- vision
- water
- water purification

REFERENCES

1. Aghabozorg, H. R., & Hassani, S. S., (2017). Removal of pollutants from the environment using sorbents and nanocatalysts, (Chapter 4), *Advanced Environmental Analysis: Application of Nanomaterials, RSC Detection Science,* Chaudhery Mustansar Hussain, Boris Kharisov (Eds.). (Royal Society of Chemistry, U.K.), 74–89.
2. Azzaza, S., Thinesh Kumar, R., Judith Vijaya, J., & Bououdina, M., (2017). Nanomaterials for heavy metal removal, (Chapter 7). *Advanced Environmental Analysts: Application of Nanomaterials, RSC Detection Science,* Chaudhery Mustansar Hussain, & Boris Kharisov (Eds.). Royal Society of Chemistry, U.K., 139–186.
3. Chakraborti, D., Das, B., Rahman, M. M., Choudhury, U., Biswas, B., Goswami, A. B., et al., (2009). Status of groundwater arsenic contamination in the state of West Bengal, India: A 20 year study report, *Mol. Nutr. Food. Res., 53,* 542–551.
4. Guha, M. D. N., (2008). Chronic arsenic toxicity and human health. *Indian Journal of Medical Research, 128,* 436–447.
5. Dey, T. K., Banerjee, P., Bakshi, M., Kar, A., & Ghosh, S., (2014). Groundwater arsenic contamination in West Bengal: Current scenario, effects and probable ways of mitigation, *International Letters of Natural Sciences, 13,* 45–58.
6. Smith, A. H., Lingas, E. O., & Rahman, M., (2000). Contamination of drinking water by arsenic in Bangladesh: A public health emergency. *Bulletin of the World Health Organization, 78*(9), 1093–1103.

7. Hashim, M. A., Mukhopadhyay, S., Sahu, J. N., & Sengupta, B., (2011). Remediation technologies for heavy metal contaminated groundwater. *Journal of Environmental Management, 92*, 2355–2388.
8. Shannon, M. A., Bohn, P. W., Elimelech, M., Georgiadis, J. G., Marinas, B. J., & Mayes, A. M., (2008). *Science and Technology for Water Purification in the Coming Decades*, Nature, Nature Publishing Group, USA, 301–310.
9. Palit, S., (2016). Nanofiltration and ultrafiltration: the next generation environmental engineering tool and a vision for the future. *International Journal of Chem. Tech. Research, 9*(5), 848–856.
10. Palit, S., (2016). Filtration: Frontiers of the engineering and science of nanofiltration- a far-reaching review. In: Ubaldo Ortiz-Mendez, Kharissova. O. V., & Kharisov. B. I., (eds.), *CRC Concise Encyclopedia of Nanotechnology (Taylor and Francis)* (pp. 205–214).
11. Palit, S., (2015). Advanced oxidation processes, nanofiltration, and application of bubble column reactor, In: Boris, I., Kharisov, O. V., Kharissova, R., & Dias, H. V., (eds.), *Nanomaterials for Environmental Protection* (pp. 207–215). Wiley, USA.

INDEX

α
α-amylase, 110, 111
α-carotene, 65, 67, 69, 70, 75
α-pinene, 133

β
β-carotene, 64–70, 72, 74–77
β-cryptoxanthin, 64, 69, 70
β-pinene, 133, 135

A
Abiotic stress, 13
Absorption, 45, 92, 172, 209
Acetaldehyde, 97
Acetate groups, 96
Acids, 5, 6, 14, 21, 22, 47, 53, 92, 97, 103, 107, 113, 115, 116, 118–120, 122–124, 133, 161, 162, 168, 191
Acinetobacter, 22
Acoustic cavitation, 76
Activated carbon, 231
Active
  coatings, 28
  compounds, 22, 103, 137, 142
  edible coatings, 18
  packaging, 17, 21, 90, 92, 132, 142
Acylglycerols, 5
Adhesion, 12
Adsorbent, 209, 212, 213, 231
Adsorption, 122, 213
Aeromonas hydrophilla, 22
Aesthetic, 3, 6, 91
Agro-industry, 108
Alcaligenes, 22
Alcohols, 123, 133
Aldehydes, 94, 133
Algae, 19, 69, 95, 200, 209, 211

Alginate, 3, 5, 12, 18, 23, 27, 44, 49, 54, 55, 102, 103, 139, 140, 199–201, 203, 205, 206, 209–212, 214, 215
  fenugreek, 44, 55
Alginic acid, 211
Aliphatic, 231
Alkaline condition, 5
Alkalis, 5
Aloe mucilage, 8
Alpha tocopherol protection, 54
Alternative materials, 55
Alveolar inflammation, 214
Ambient temperature, 12
Amines, 133
Amino acid, 4, 19
Amphiphilic nature, 4
Amylopectin, 95, 96
Amylose, 95
Analytical test, 12
Anatomic analyses, 188
Anethole, 133
Animal
  collagen, 97
  proteins, 7, 9
  slaughtering, 7
Anthocyanidins, 107, 113–115
Anthocyanins, 114, 121, 124
Anthropocene, 213
Antibacterial activities, 12
Antifungal activity, 11, 139
Antimicrobial, 2, 5, 6, 9, 10, 15–18, 21–23, 25, 26, 28, 89–92, 97, 99, 101, 103, 131–137, 139, 142, 170, 205, 209, 212
  activity, 25, 103
  compound, 22
Anti-nutritional compounds, 109, 112
Antioxidant, 2, 5, 6, 8–10, 12, 17, 20–23, 28, 63, 64, 66–68, 89, 91, 97, 101, 107, 116, 124, 131, 137, 138, 142, 143, 159, 160, 162, 168, 170, 208

activity, 8, 12, 63, 170
agents, 2
Anti-tumor immunity, 48
Apoptosis, 48
Appetite regulation, 47
Aqueous solution, 97, 204, 207, 211
Arabic gum, 6, 12
Arabinoxylan, 45
Arcobacter butzleri, 22
Aroma, 7, 8, 17, 20, 160
aromatic compounds, 110, 114, 115, 123, 133, 200, 231
deterioration, 20
Arsenic, 221–228, 230–237, 239–241
groundwater, 221–223, 233, 235
Arsenicosis, 234, 235
Ascorbic acid, 8, 18, 90, 119, 152, 153
Aspergillus, 22, 120, 123, 170, 209, 213
Assessment of persian lime quality, 11
Astaxanthin, 64, 66
Auricularia, 111
Autonomic imbalance, 214

B

Bacillus cereus, 22
Bacteria, 19, 21, 22, 44, 45, 47–50, 53–56, 65, 69, 91, 97, 99, 101, 132, 133, 139, 167, 191, 200, 209, 212
Bacterial
endophthalmitis, 210
infection, 48
Bacteriocins, 56
Barrier, 2, 4, 6, 7, 9, 13, 14, 17–20, 28, 47, 50, 91, 95, 100–103, 132, 137, 140
properties, 4, 14, 18, 19, 28, 100, 102, 137, 140
Beneficial microflora, 48
Benzoic acid, 19
Beta-cyclodextrin, 18
Bifidobacterial populations, 47
Bifidobacterium (BF), 46, 48, 49, 53–55
bifidum, 46, 55
lactis BB12 (L), 46
longum BB 536 (L), 46
Bilateral symmetry, 65
Bioactive

components, 6, 52, 135
compounds, 9, 19, 43, 50, 52, 64, 76, 101, 124, 160, 162, 163, 166–168, 170–172, 202
ingredients, 44, 49
proteins, 95
substances, 47, 90, 92, 93
Bioavailability, 68, 172, 214
Biochothrix thermospacta, 23
Biocompatibility, 9, 44, 96, 202, 212–214
Biocompatible, 51, 93, 95, 99, 103, 200, 214
Biodegradability, 7, 9, 44, 96, 99, 101, 213, 214
Biodegradable, 3–6, 9, 11, 17, 24, 28, 91, 93, 95–97, 99–104, 131, 200, 212, 214
Bioethanol, 110, 111, 205
Biofertilizer, 110
Biofunctionality, 51
Biogas, 110
Biological systems, 214
Biomacromolecules, 94, 101
Biomaterials, 97
Biomolecules, 94, 95, 101, 170
Biopolymers, 4, 6–8, 15, 22, 52, 91, 96, 100–103, 105, 199–201, 203, 205, 208, 209, 212–215
Biopreservation technology, 16
Biosorption, 236
Biosynthesis, 72, 73
Biosynthetic genes, 69
Biotechnology, 77, 96, 99
Biotransformation, 107–109, 111, 170
Bitter vetch protein, 9
Bladder, 68, 234
Blakeslea trispora, 65
Bovine
hides, 19
serum albumin (BSA), 213
spongiform encephalopathy, 7
Breba figs, 8
Briquettes, 110
Brochothrix thermosphacta, 22, 136
Butylated hydroxytoluene (BHT), 160
Butyrate, 48, 49

Index

C

C. arabica, 108, 117
C. canephora, 108, 117, 121
C. jambhiri, 188
Cactaceae, 149, 150, 152, 186
Caffeic, 116, 119, 120, 122–124
  acid phenethyl ester (CAPE), 123, 124
Calcium, 64, 90, 99, 102, 211
Campylobacter spp., 22
Cancer, 48, 63, 64, 68, 201, 205, 206, 211, 212, 214, 234, 235
  cells, 48, 212, 214
Candelilla, 12, 13
Candida strains, 22
Capsanthin, 64, 70
Capsicum anuum, 70
Capsorrubina, 70
Carbohydrates, 5, 6, 51, 56, 112, 153, 161
Carbon, 2, 63, 65, 72, 76, 78, 110, 114, 201, 231
  dioxide, 2, 12, 76
  nanotubes, 231
Carboxyl, 65, 115, 122, 200, 207
Carboxylate groups, 99
Carboxymethyl
  cellulose, 5, 15
  starch, 15
Carbures, 133
Carcinogenic, 48, 91
Carcinogens, 47, 48
Cardiovascular
  diseases, 63, 64, 68, 79
  health, 66
Carica papaya, 70
Carnauba, 13
  wax, 12
Carotenoid, 8, 18, 63–79, 161, 168
  profile, 69
  recovery, 74
  triplet, 72
Carrageenan, 5, 18, 21, 27, 44, 51, 138, 199–201, 203, 206, 209, 212, 213, 215
Carvacrol, 23, 27, 133–135, 140
Casein, 7, 52
Caseinates, 102

Cassava, 96, 139
Catastrophe, 224, 228, 230, 235, 236, 238–240
Catechin, 8, 113, 114, 161, 162, 165, 168, 169
Cation, 99
Cavitation, 167
Cell
  cytotoxicity, 68
  death, 72, 133, 210
Cellular
  changes, 69
  surfaces, 99
Cellulose, 5, 6, 8, 15, 18, 51, 95, 103, 112, 136, 199, 200, 203, 205, 207, 208, 215
  fiber, 207
  nanocrystals, 205, 207
  nanofibers, 207
  nanoparticles, 205, 207
Center and Development for Food Industries (CIDIA), 1, 159
Ceratocytis fimbriata, 111
Cereals, 193
Cervical, 68
Chelating, 24
Chemical
  cross-linking, 5, 92
  hydrogels, 94, 96
  hydrolysis, 5, 119, 203
Chicory, 49, 53, 54
Chitosan, 5, 8, 10, 11, 15, 16, 18, 21, 23, 25, 27, 44, 51, 54, 55, 91, 98, 99, 102, 103, 138, 139, 143, 199, 200, 203–205, 209–211, 213–215
Chlorella vulgaris, 77
Chlorogenate hydrolase, 110, 111, 123
Chlorogenic acid, 113, 116, 119–123
Chloroplasts, 69
Chronic, 234, 235
  arsenic toxicity, 234
  intake, 235
Cinnamaldehyde, 23, 102
Cinnamomum zeylanicum, 102
Cinnamon, 21, 23, 101, 134, 136, 139
Cinnamyl alcohol, 133
Cis-trans isomers, 66

Citral, 18
Citric acid, 18, 92, 153
Citronellol, 133
Citrus, 2, 70, 102, 134, 135, 182, 187, 188, 193
Clostridium
 botulinum, 22
 perfringens, 22
Coacervation, 50
Coal dewatering, 95
Coated guava, 12
Coating, 2–7, 9–11, 13–22, 24–28, 91, 99, 101, 102, 201, 203, 205, 208
 formulation, 4, 12, 17, 19
 functions, 6
Cocrystallization, 50
Co-encapsulation, 55
Coffee
 phenolic compounds, 109
 pulp (CP), 107–124
Cohesive biomaterials, 5
Collagen, 7, 10, 15, 19, 20, 91, 97
Colloidal stability, 103
Colon, 45, 48, 68
 cancer, 48
 microflora, 55
Colonic glutathione S-transferase, 48
Colonization, 47, 48
Color analysis, 11, 25
Colorants, 5, 6, 63
Colossus, 229, 230, 234
Commercialization, 13, 52, 91, 132, 153
Complexation, 201, 209, 211, 213
Compounds, 5, 7, 10, 15, 18, 22, 25, 27, 28, 50, 51, 56, 63, 64, 70, 72, 75, 76, 78, 79, 93, 101, 108, 109, 111, 113–118, 121, 122, 124, 133, 136, 137, 142, 159, 160, 162–172, 207, 231
Contamination, 11, 26, 100, 222–225, 227, 228, 230–233, 235, 236, 239, 240
Conventional oral iron therapy, 211
Copolymer, 96, 136
Core-shell structure, 211
Corn zein, 7
Cornerstones, 227, 229, 231, 236
Coronary heart disease, 89, 90

Corrosion, 100
Corynebacterium, 22
Cosmetic
 alimentary, 63
 industry, 66, 68
Cost-effectiveness, 44
Covalent, 92, 97, 98, 120, 209
Crops homogeneity, 191
Cross-linking, 11, 92–94
Crucial juncture, 223, 228
Crude fiber, 112
Cryptococus, 65
Crystalline
 character, 97
 nuclei, 97
 solid states, 5
 structure, 207
Cultivars, 117, 187, 190, 191, 193
Curcumin, 212
Cyclodextrins, 55
Cymene, 133, 140
Cytotoxic potential, 212
Cytotoxicity, 214

D

Daptomycin, 210
Deboning, 90
Decontamination, 12, 227, 237
Delineates, 223, 230, 234
Denaturation, 19, 73
Dendritic cells, 48
Derivatives, 4, 90, 107, 113–116, 118, 119, 121, 124, 133, 168
Deterioration, 14, 17, 20–22, 44, 89–91, 97, 101, 142
Detoxification, 48, 109, 202
Diabetes, 64, 67, 68
Diarrhea, 48
Dietary supplements, 68
Diffusion, 15, 17, 49, 75, 136, 142, 168
Digestive enzymes, 51
Dipping, 18
Discoloration, 8, 22
Dispersibility, 212
Dissolution, 92, 94, 103, 204, 207, 210
Distillation, 132, 162

Index

Domain, 93, 94, 224, 227, 231, 234
Double bond, 64–66, 123
Drug delivery systems, 95, 214
Dry jet-wet spinning, 207

E

E. fetida, 110
Ecological biodiversity, 228, 229
Edibility, 7
Edible
  casings, 103
  coating, 1, 3, 4, 10–13, 15, 17–19, 23, 24, 26, 28, 91, 102, 104
  films, 2–15, 17, 22, 24–28, 141, 143
Eisenia andrei, 110
Elasticity, 7, 92, 140
Electric pulses, 74
Electrospinning, 56
Electrostatic
  charges, 4
  forces, 56
  interaction, 214
Elongation, 9, 10, 140
Emancipation, 221–226, 229, 231, 239
Embryogenesis, 188
Emission, 10, 69, 167, 231
Emitters, 21
Emulsion, 2, 6, 12, 14, 140, 201
Encapsulated, 44, 49–52, 56, 139
  probiotic, 55
Encapsulating agents, 44, 51
Encapsulation, 43–45, 49–51, 53, 55, 56, 210
  processes, 49
  systems, 44
Endeavor, 221–225, 227, 228, 231, 232, 234–238
Endocrinological effects, 47
Endogenous enzymes, 20
Endometrial, 68
Energy
  crisis, 229
  intake, 47, 48
  security, 229
  sustainability, 228–230
Enigma, 226, 232, 238

Enigmatic issue, 223, 226, 232, 236
Enrobed fish sticks, 21
Enterobacter, 22
Enterobacteria, 10, 25
Enterobacteriaceae, 8, 136
Enterococcus
  fecium, 49, 54
  SF68, 46
Enterocytes, 47
Enterohemorrhagic *E. coli* (EHEC), 22
Environmental
  crisis, 221, 224, 228, 230
  disaster, 222, 225, 238
Enzymatic
  activity, 14, 47
  lysis, 74
  modification, 108, 109
  reactions, 21, 72, 124
  treatment, 5
Enzyme, 21, 73, 90, 93, 110, 120, 161, 209, 210, 212
Epicatechin (EC), 10, 114, 115, 164, 165, 168, 169
  gallate (ECG), 164, 165, 168, 169
Epidemiological studies, 64, 68, 116
Epigallocatechin (EGC), 164, 165, 168, 169
  gallate (EGCG), 164, 165, 168, 169
Epithelial
  defensin production, 48
  tight junctions, 48
Escherichia coli, 22, 46, 136
Essential oils (EOs), 21, 22, 97, 131–137, 140, 142, 144
Esterase, 110, 111, 119, 120, 123
Esterified, 116, 120, 123
Esters, 10, 116, 119, 123, 133
Ethanol, 27, 75, 79, 111, 118, 121, 163, 166, 200, 207, 210
Ethylene, 2, 12, 16, 89, 91, 136
  vinyl alcohol (EVOH), 136
Eugenol, 18, 134
Extraction, 25, 74–77, 107–109, 111, 112, 117–120, 124, 134, 159, 160, 162, 163, 166–171, 206, 212
Extrusion, 24, 50, 207

F

F. oxysporum, 213
Fabrication, 102
Fatty acid esters, 5
Fermentable, 111
Fermentation, 44, 47, 72, 78, 91, 108, 109, 111, 120, 124, 161, 162, 168, 170
Fermentative route, 70
Ferrous, 211
Ferulic acid, 11, 113, 119, 120, 123
Feruloyl, 110, 111, 113, 116, 119, 123
Fibers, 5, 56, 103, 203, 207
Film, 2–7, 9, 12–17, 19–26, 28, 90, 93, 101, 102, 136–140, 143, 144, 207, 212, 213
  conditioning, 10
  forming, 11, 19
    materials, 4, 5
  integrity, 4
  network, 7
  solubility, 12
  thickness, 10, 12
  transparency, 12
Firmness, 8, 11, 12, 18, 90, 91, 102
Flavanols, 107, 113–115, 118, 168
Flavobacterium, 22
Flavonoids, 114, 117, 118, 122, 124, 160, 165, 169
Flavonols, 107, 113–115, 162, 168
Flavor, 5, 6, 17, 142, 155
  releasers, 21
Flexibility, 5, 7, 9, 19, 93, 94
Flow injection synthesis, 203
Fluconazole, 213
Fluidization, 24
Folin-ciocalteu, 117
Food,
  additives, 21, 90, 93, 95, 131
  development, 149, 156
  Drug Administration (FDA), 26, 133, 167
  industry, 4, 26, 44, 45, 50, 52, 53, 55, 56, 66, 68, 77, 99, 116, 132, 171, 182, 185, 187, 192, 193, 199, 200, 202, 205, 206, 209, 210, 212, 215
  packaging, 6, 28, 100, 132, 142
  quality, 2, 4, 28
  sensory quality, 51
Formaldehyde, 97, 208
Formulate active packaging, 23
Fragile, 4
Freeze
  drying, 50
  thaw cycles, 92
Frequency, 160, 167, 170, 188, 189, 207
Fresh cut fruits (FFC), 90, 91
Frozen meat, 3
Fructan, 48
  prebiotics, 47
Fructooligosaccharides (FOS), 45, 49, 54
Functional
  foods, 43–45, 50, 56, 149
  ingredients, 2, 44
Functionalization, 202, 207
Fungal
  growth, 25
  infections, 15
Fungi, 19, 65, 69, 72, 97, 99, 101, 109, 111, 133, 209
Fungicides, 3, 102
Fusarium, 22, 139

G

Galactonannans, 10
Galactooligosaccharides (GOS), 45, 54
Gallic, 116, 118, 120, 168
Gallocatechin (GC), 10, 161, 168
  gallate (GCG), 168
Gamma radiation, 3
Gangrene, 235
Gas transfer rates, 10
Gastric damage, 68
Gastrointestinal
  conditions, 50
  diseases, 48
  microbiota, 45
  tract, 45, 46, 50, 53, 55
  transit, 47
  zone, 43

Gelatin, 2, 7, 8, 10, 12, 15, 19–21, 23, 25, 27, 51, 52, 91, 97, 98, 102, 103, 139, 140, 143
  coatings, 19
  pectin, 102
  types A and B, 97
Generally regarded as safe (GRAS), 6, 55, 133
Generic structures, 114
Genes, 69, 188
Genetic improvement, 191
Genotoxic compounds, 48
Genotype, 69, 189
Geology, 225, 233
Geraniol, 133
Geranyl-geranyl diphosphate, 65
Germplasm, 188, 189
Ghatti gum, 13
Glacial acetic acid, 8
Glucose, 72, 78, 79, 95, 116, 168, 207
Glucosides, 114
Glutaraldehyde, 97, 209, 210
Gluten, 7, 9, 15
Glycerol, 8, 10–12, 15, 18, 19, 25, 27, 72, 123
Golden apple, 193
Graphenes, 231
Grave, 224, 225, 236, 237, 240, 241
Grinding, 203
Groundbreaking, 232, 239, 240
Groundwater, 221–241
  poisoning, 223–227, 235, 236
  remediation, 222, 223, 225–227, 232, 234, 236, 239, 240
Gum, 3, 5
  locust bean gum matrix, 55
  tragacanth, 13
Gumminess, 21
Gut hormones, 47

H

Hallmarks, 226, 229, 238
Handling, 4, 5, 69, 102, 110, 163
Hazelnut meal protein, 9
Health hazards, 163, 223, 234, 235

Heavy metal, 209, 221, 222, 224–228, 230–232, 235–237, 239
Helicobacter pylori, 48
Hepatoprotective properties, 116
Herbivores, 132
Heredity, 188, 189
Heterogeneous structure, 22
Hexane, 75, 76, 118, 163
Hexyl acetate, 10
Hidroxypropyl methylcellulose, 12
High
  performance liquid chromatography (HPLC), 77, 114, 115, 122
  speed counter-current chromatography (HSCCC), 122
Hipotiocianita analysis, 25
Homeostasis, 191
Homogeneous distribution, 50
Hormones, 47, 66
Human
  civilization, 221–233, 235, 236, 239, 241
  scientific endeavor, 225
Humidity, 7, 16, 91, 110, 184
Hydrocarbon, 64, 65
  carotenes, 64
  compounds, 65
Hydrochloric acid, 208, 211
Hydrodistillation, 132
Hydrogel, 89, 90, 92–102, 104, 105, 207, 213
  structure, 92
  variants, 102
Hydrogen, 24, 65, 90, 92, 94, 97, 98, 208
Hydrogenated fats, 51
Hydrolysis, 45, 96, 97, 111, 118, 119, 124, 163, 210
Hydrolyzable, 116, 165
  tannins, 116
Hydrolyzation, 19
Hydrophilic
  character, 28
  exterior, 55
  functional groups, 94
  head, 52
  moieties, 5

motifs, 52
Hydrophobic
  acyl chain, 52
  components, 15
  interior, 55
  materials, 14
  nature, 92
Hydrothermal models, 203
Hydroxybenzoic acid, 115, 116
Hydroxycinnamic acids, 113, 115, 116, 118, 119
Hydroxyl, 5, 65, 66, 114, 115, 122, 200, 207, 209
  groups, 5, 207, 209
Hydroxymethyl starch, 15
Hydroxymethylation, 208
Hydroxypropyl starch, 15
Hypercholesterolemia, 89, 90
Hyperpigmentation, 235

I

Illumination, 72
Immense
  lucidity, 233, 235
  scientific
    girth, 233
    prowess, 223–225, 229, 234, 238, 239
    regeneration, 230
Immune function, 66, 67
Immunological effects, 47
Immunomodulation properties, 48
Immunomodulatory capacity, 48
Implementation, 28, 155
In vitro, 68, 95, 116, 138, 188, 205
In vivo, 67, 68, 95, 116
Indo-Bangladesh Gangetic delta, 235
Inflammation, 68, 159
Ingenuity, 222, 223, 225, 229, 234, 236, 237, 240
Inhibition, 8, 12, 78, 123
Inhibitory concentration, 25
Inorganic fertilizer, 110
Integrity, 5, 21, 43, 68, 91, 93, 142, 172
Intensity, 26, 72, 77, 78, 167, 204
Interchain interactions, 101
Intermittent microwave radiation, 76

Internal network structure, 92
International Scientific Association of Probiotics and Prebiotics (ISAPP), 45
Interpenetrated network hydrogels (IPN), 92
Intestinal
  microbiota, 44
  zone, 49
Intricacies, 221, 223, 240
Intrinsic probiotic factors, 46
Introspection, 223, 224, 226, 228, 230, 231, 237, 239
Inulin, 45, 47–49, 53, 54
Inverted sugar, 25
Iodide, 8
Ionic gelation, 201, 210, 211
  method, 211
Ionized carboxylate groups, 99
Ionotropic gelation, 51
Iron encapsulation, 211
Irradiation, 4, 97, 204
Isoelectric point, 11, 213
Isomerization, 65, 66

K

Kaempferol, 168, 211
Karaya gum, 13
K-carrageenan gum, 18
Keratosis, 234, 235
Ketones, 133
Klebsiella, 22

L

Laccase, 110, 111
Lactic acid bacteria, 55
Lactobacillus, 22, 44, 46, 48, 49, 53–55
  acidophilus LA1, 46
  casei (L), 46, 49, 54, 55
  plantarum, 44, 46, 49
  reuteri, 46, 48, 55
  rhamnosus GG, 46
  sporogens, 46
Lactoperoxidase system, 25
Lactulose, 45, 48
Landscape, 230

Larding, 3
Larrea tridentata, 77
Laurel, 10, 23
Lentinula, 111
Lethal effects, 69
Leucoanthocyanidins, 114
Leuconostoc spp., 22
Light harvesting, 66, 69
Lignin, 21, 112, 161, 199, 200, 203–205, 208, 215
Limonene, 18, 134, 135
Linear
 low-density polyethylene (LLDPE), 136
 polyanions, 99
Linseed, 8
Lipid, 4–6, 9, 12–14, 17, 25, 27, 51, 52, 91, 96, 121, 133, 135, 137, 140, 142, 161
 membranes, 69
 oxidation, 8, 21–23, 25, 27, 137
 solubility, 2
Lipidic moieties, 52
Lipophilic
 nature, 63
 pigment, 64
Liquid
 extraction, 75, 118, 162, 168
 fraction, 101
 nuclear wastes, 231
Listeria monocytogenes, 22, 23, 27, 136–138, 212
Low-methoxyl (LM), 8, 27, 99
Lucidly, 222, 231
Lutein, 66–70
Lycopene, 64–66, 68–70, 76–78
Lycopersicum esculentum, 70
Lyocell process, 207
Lysine, 189, 191
Lysozyme, 10, 22, 205, 210

M

Maceration, 74, 132, 162
Macromolecular chains, 93
Macromolecules, 4, 94, 116, 209
Macular degeneration (AMD), 63, 67
Magnesium, 64, 168

Magnetic resonance imaging (MRI), 202, 205
Maize, 44, 95, 136, 188–191, 193
Maleic acid, 92
Mammalian gelatins, 9
Mango, 15, 96, 189, 191, 193
Marauding domain, 224
Mastic resin, 13
Matrix, 9, 44, 50, 53, 56, 75, 76, 79, 92, 97, 137–140, 163, 166, 200, 212
Maturation, 12, 69, 90, 102
Meatballs, 10
Membranes, 101, 103, 167, 214
Mesquite gum, 13, 103
Metabolism, 46, 47, 49, 72
Metabolization, 46
Metal
 ions, 5, 24, 231
 oxides, 202, 231
Methanol, 97, 118, 121, 163, 166
Methotrexate, 212
Methoxy, 65, 134, 200
Microbial
 cells, 71, 133
 contamination, 16, 28, 137
 count, 23
 degradation, 21
 growth, 19–22, 25, 26, 73, 74, 112, 138
 membrane, 133
 pigments, 77
 production, 70, 77, 79
 spoilage, 6, 16, 18, 20, 21
Microbiological
 analysis, 27
 safety, 16
Microbiology, 8, 12, 25
Microbiota, 46, 47, 49
Microcapsules, 44, 51, 53
Microencapsulation technique, 53
Microfibrillated, 205, 207
Micrometric size, 49
Microorganisms, 8, 22, 25, 26, 44, 46, 49–51, 63, 64, 71–74, 77–79, 89, 90, 96, 100, 111, 120, 132, 136, 139, 142, 162, 170, 209
Microwave, 76, 118, 121

assisted extraction (MAE), 118, 121, 166, 169
Migration, 5, 19, 65, 135
Milling, 203, 204
Mitigation, 235
Modified atmosphere packaging (MAP), 90, 136
Moisture, 2–4, 9, 10, 13, 17, 19, 21, 24, 25, 50, 78, 100, 109, 111, 120, 138, 140
  barrier, 2, 9, 100
  contents, 10, 25
  loss, 2, 13, 25
Mold growth, 16
Molecular
  identification, 77
  mobility, 5
  structure, 4, 64
  weight, 5, 15, 19, 117, 200, 209
Molecule, 51, 52, 63, 65, 74, 93, 99, 114, 166, 201, 202, 212
  backbone, 101
Molten wax coating, 2
Monilia, 22
Monoaldehydes, 97
Monocyclic, 133
Monoterpenes, 133
Montmorillonite, 25
Moraxella, 22
Morphology, 150, 189, 204, 207
Mucilage solution, 8
Multicomponent systems, 94
Multiple sclerosis, 63
Multivalent cations, 5
Muscle foods, 17, 21–23
Mutagenic, 48
Mycobacterium spp., 22

N

Nano-aggregates, 211
Nanocarriers, 214
Nanocatalysts, 231
Nanocomposite, 25, 207, 212, 213, 231
  coating, 25
Nanocrystals, 203, 205, 207
Nanoemulsion, 12
Nano-engineering, 231

Nanofillers, 213
Nano-intermediates, 201
Nanomaterials, 56, 201, 202, 214, 230–232, 238, 241
Nanometric materials, 214
Nanoparticles (NPs), 52, 101, 201–215
Nano-range, 56
Nanoscale, 56, 201, 211
Nano-science, 231
Nanosized, 231
Nanostructured, 207
Nanostructures, 199, 201–203, 205, 213, 215
Nanosystems, 201, 203, 213, 214
Nanotechnological use, 214
Nanotechnology, 201, 202, 213, 215, 230–232, 238
Nanotube, 201
National Council of Science and Technology of Mexico (CONACyT), 193
Natural
  colorants, 63
  killer cells, 48
  materials, 6
  matrix, 66
  plasticizers, 5
  polymers, 6, 215
Neoplastic changes, 48
Network chains, 94
Neuroblastoma, 67
Neurological effects, 47
Neuronal differentiation, 67
Neutral, 5, 168
  carbohydrate, 5
Nitrogen, 79, 112, 119, 153
  source, 79
Noncarbohydrate compounds, 45
Non-encapsulated
  bacteria, 44
  probiotics, 51
Non-toxic, 51, 95, 99, 132, 199, 200, 209
Nontoxicity, 2, 44
Nopal cactus, 8
Nozzles, 26
Nucellar
  apomixis, 187

embryos, 186, 191
plantlets, 189
plants, 191
science, 226
Nutraceuticals, 5
Nutrients, 13, 14, 17, 20, 109, 110
Nutritional
supplement, 79
value, 7, 9, 66, 102, 112, 150

O

Obesity, 47, 89, 90
Ocular epithelia, 210
Odor, 16, 21, 25
Ohmic heating, 78
Oil diffusion, 17
Olea europaea L., 190
Oligofructose (OF), 47
Oligomeric, 114, 117
Oligosaccharides, 45, 54, 123
Olive, 13, 138, 190, 193
Opaque, 4
Optimum conditions, 77, 122
Opuntia ficus-indica cladodes, 8
Oral
cavity, 68
ingestion, 101
Oregano, 12, 13, 19, 23, 134, 136, 138–140, 143
essential oil, 12, 13, 136, 140
Organic
compounds, 202, 212, 231
fertilizer, 91, 110
solvents, 75, 121, 209
Organisms, 46, 56, 69, 91, 99, 182
Organoleptic, 1, 6, 23, 154, 168
Origanum virens, 25
Oryza sativa L., 190
Oscillatory movement, 233
Osmotic pressure, 51
Oxidation, 6, 17, 21, 22, 24, 25, 27, 28, 66, 69, 119, 142, 161, 162, 227, 240
Oxidative
damage, 64, 68
stress, 67

Oxygen, 2, 7, 12, 15, 19, 22, 50, 65–70, 73, 74, 91, 100–102
Ozone, 90

P

P. pinophilum, 77
P. purpurogenum, 77, 78
Packaging, 4–7, 9, 16, 17, 20–22, 25, 26, 28, 52, 89–93, 97, 99–103, 131–133, 135, 136, 139, 142, 144, 201, 202, 210
Paclitaxel (PTX), 211, 212
Pallbearers, 235
Pancreas, 68
Paradigm, 225, 226, 234, 238, 240
Parameter, 73, 140, 227
Partial alkaline hydrolysis, 97
Path-breaking, 226, 230
Pathogens, 12, 26, 50, 132, 133, 210
P-coumaric acid, 113, 116, 119, 120, 168
Pectin, 5, 6, 8, 10, 18, 44, 45, 51, 99, 102, 103, 166, 168
Pectinase, 110, 111, 120
Peeling, 16, 26, 89, 90
Pellets, 110
Pencillin sp., 213
Penetration, 75, 76, 166, 167
Penicillium purpurogenum, 77, 120
Perfumery, 133
Pericarp, 109, 150
Perionyx excavates, 110
Perishable, 13, 20, 24, 155
Permeability, 6, 10, 15, 17, 19, 25, 52, 133, 137, 138, 140, 167
Permeation, 5, 210
Peroxide, 10, 90, 143, 205, 208
Petroleum, 3, 4, 75, 121, 230, 231
pH, 8, 10–12, 19, 20, 25, 27, 43, 49, 51, 55, 66, 74, 78, 91, 93, 97, 99, 101, 120, 152, 153, 208, 209, 213
Pharmaceutical, 66, 95, 170
Phenol, 23, 133, 134
Phenolic
acids, 115, 119, 120
compound, 97, 101, 107–122, 124, 140, 142, 159
type, 133

Phosphorus, 64, 168
Photochemiluminescence (PCL), 143
Photogenic microorganisms, 26
Photooxidations, 208
Photoprotectors, 69
Photosensitizers, 72
Photostability, 212
Photosynthetic apparatus, 69
Phototrophic bacteria, 69
Physicochemical
　characteristics, 52, 95
　parameters, 8, 120
　properties, 27, 51, 52
Physiological fluid, 92
Phytochemical, 13, 161
　compounds, 149
Phytoplasma, 191
Pigments, 64, 69, 71, 74, 75, 77, 78, 121
　production, 77, 78
　protein complexes, 69
　recovery, 74
Pigskins, 19
Pitahalla, 150
Pitajalla, 150
Pitajaya, 150
Pitalla, 150
Pitaya fruit, 156
Pivotal importance, 222
Plants, 9, 63, 64, 69, 71, 95, 108, 110, 116,
　132, 150, 159, 160, 162, 163, 181, 182,
　184–186, 188, 191–193, 199, 213
　extracts, 22
　matrix, 75, 166
Plantago psyllium, 49
Plasma vacuoles, 69
Plasticizer, 5, 7, 9, 14, 15, 19
Pleorotus, 111
　ostreatus, 53
Poli(vinyl alcohol) (PCA), 96
Pollutants, 167, 231
Poly(lactic-co-glycolic acid) (PLGA), 211
Polycarboxylates, 99
Polycarboxylic acids, 101
Polyelectrolyte multilayer, 8
Polyembryonic, 185, 188–191
Polyembryony, 181–193

Polyester, 4
Polyethylene glycol, 18, 19, 51, 212
Polyfunctional monomers, 92
Polyhydroxyalkanoates (PHA), 91
Poly-isoprenoid structure, 65
Polylactic acid (PLA), 91
Polymer, 1–7, 9, 14, 16, 21, 22, 44, 49, 52,
　56, 92–101, 117, 119, 136, 140, 200,
　202, 208, 211, 212, 215
　chains, 94, 97
　molecules, 5
Polymeric
　applications, 208
　backbone, 94
　matrix, 5, 207
　mixture, 55
　packaging, 6
　plastics, 4
　proanthocyanidins, 114
　solids, 93
　structures, 90, 93
Polymerization, 24, 92, 96
Polyol, 116, 203, 204
　methods, 203
Polyphenolic compounds, 114
Polyphenols, 23, 45, 64, 107, 120–124,
　161–163, 168, 172
Polypropylene, 136
Polypyrrole, 206, 212
Polysaccharides, 4–6, 8, 9, 14, 15, 19, 44,
　51, 52, 91, 95, 102, 120, 137, 168, 200,
　201, 212
Polyurethane foam, 78
Polyvinyl alcohol (PVA), 96–98, 101, 103,
　136
Porter's reagent method, 117
Postharvest protection analyses, 18
Post-mortem storage conditions, 20
Potassium sorbate, 27
Potential, 9, 12, 28, 55, 56, 92, 99, 100,
　103, 109, 112, 116, 123, 149, 154, 155,
　160, 163, 182, 187, 188, 192, 193, 201,
　202, 205, 208–210, 212, 223, 239
Prebiotic, 43–50, 52–56
　compounds, 44
　fermentation, 48

Pressurized hot water extraction (PHWE), 167
Proanthocyanidins, 113, 115, 117, 118, 121, 124
Probiotic, 43–56
Profundity, 222, 223, 227–229
Prolongation, 14, 28
Prostate, 68
Proteinaceous, 2
Proteinantioxidant interactions, 22
Proteinpectic coating, 18
Proteins, 4–7, 9–11, 14, 15, 17, 19, 20, 51, 52, 56, 91, 96, 97, 103, 112, 116, 134, 137, 153, 165, 201, 214
Proteus spp., 22
Protocatechuic acid, 115
Provision, 222, 227, 229, 238
Provitamin A, 65, 67
Prunus persica, 70
Pseudomonas, 22, 23, 136
Psidium guajava, 70
Pullulan, 5, 8
Pullulanase, 209, 212
Pulmonary, 214
Pulp, 107–110, 117, 121, 124, 150, 152, 207
Purge accumulation, 20
Purification, 107–109, 117, 121, 122, 209, 211, 227, 237

Q

Qualitative alterations, 48
Quality parameters, 12
Quercus sp, 77
Queretaroensis, 149, 150, 152–154
Quinoa, 11
  protein, 9, 16

R

Radicles, 189
Rancidity, 2, 4, 20
Rayon, 207
Reactive groups, 93
Reclamation, 237
Recycling, 91, 100

Red crimson, 8
Reduction, 16, 23, 49, 67, 75, 137, 204, 237
Refrigeration, 3, 21, 90
Remediation, 222–227, 229–241
Replete, 225, 232, 234
Reproducibility, 212
Resins, 5, 6, 9, 13
Resistance, 5, 7, 16, 43, 47, 51, 55, 94, 100, 133, 163
Respiration, 2, 12–16, 21, 90, 91
  rate, 2, 12, 14, 15, 21, 91
Respiratory processes, 133
Revamped, 226, 228–230, 232, 238
Rheological
  behavior, 207
  characteristics, 2, 17
  properties, 5, 205
Rheology, 24, 28
Rhizopus, 22
  stolonifer, 12
Rhodophyceae, 201
Rhodosporidium, 65
Rhodotorula, 65, 72, 73, 77, 78
Rice, 13, 72, 95, 190, 193
Rigidity, 97, 103, 200
Ripening, 18, 69, 108, 150
  effects, 18
Rootstock, 188, 191
Roseburia sp., 49
Rosemary, 19, 23, 134
Rutin, 114, 120

S

Sabinene, 133, 134
Saccharomyces boulardii, 46
Salmon, 20, 21, 23, 66
  fillets, 21
Salmonella infections, 211
Salmonellosis, 25
Sanitation, 237
Scavengers, 21
Scientific
  barriers, 221, 222, 227, 229, 238, 240
  discernment, 224, 226, 229, 234, 239
  divination, 224, 238, 240

failures, 225
firmament, 227, 230, 232, 239, 240
fortitude, 222–226, 232, 237, 239
innovation, 222, 223
passion, 225
profundity, 226, 231, 234, 235
rejuvenation, 229, 232, 234, 237, 239
vision, 221–223, 226–230, 237, 239, 240
Seedlings abnormality, 188
Self-fertilizing mulching biopolymers, 103
Semidesert environments, 150
Sensorial properties, 27, 52
Sensory
  characteristics, 19, 102, 137
  evaluation, 8, 10, 11, 25, 27
  scores, 20, 54
Shelf life, 2–4, 8–18, 20–25, 28, 90, 91, 131–133, 136–138, 142, 154
Shellac, 12, 13
Shrimp muscle proteins, 7
Sikkim mandarin, 188
Sodium
  caseinate, 12, 51
  dodecyl sulfate, 12
Soft-solid, 5
Sol-gel, 203
Solid
  fuel, 110
  matrix, 75
  state fermentation (SSF), 78, 109, 111, 120, 170
Soluble solids, 8, 11, 152
Sorbents, 231
Sorbitol, 8, 10, 19
Soxhlet, 75, 76, 162, 165
  extraction, 75
  process, 75
Soy protein, 7, 9, 11, 15, 18
Spheroidal nanoparticles, 210
Spinacia oleracea, 70
Spinning atomizer, 50
Spoilage, 12, 20, 22, 26
Spondias lutea, 70
Sporidiobolus, 65
Sporobolomyces, 65

Sporotichum, 22
Spray
  chilling, 50
  drying, 50
Stabilization, 43, 103, 205, 208
Stabilizers, 90
Stabilizing, 48, 201, 212
Staphylococcus aureus, 12, 22, 136
Starch, 5, 6, 11, 15, 25, 44, 49, 54, 79, 95–97, 101, 139, 140, 206, 212
Steam explosion, 74
Stem distillation, 132
Stenocereus, 149, 150, 152–156
  queretaroensis, 149, 150, 152, 156
Sthapylococcus succinus, 49
Stimuli, 93, 101
Storange stability, 25
Subjective quality analysis, 12
Substrate, 45, 72, 73, 78, 111, 124
Sucrose, 2, 25
Sulfides, 133
Sulfur compounds, 231
Supercritical fluids, 74–76, 163, 166
Superparamagnetic, 210, 211
  alginate nanoparticles, 211
Surface
  area, 26, 56, 76, 167, 214
  charge, 207
  tension, 12
Sustainability, 152, 214, 223, 224, 227–230, 232, 236, 238–240
Swelling, 94, 97, 101, 213
Synthetic polymers, 6, 93, 95, 102, 104, 214

T

T regulatory cells, 48
Tackling, 222, 226, 228, 232, 235
Tannase, 110, 111
Tannins, 112, 114, 116–118
Tea, 18, 23, 25, 159–162, 164, 165, 169–172
  polyphenols, 23
Technological validation, 226
Tensile strength, 10, 101, 103, 140
Terpenes, 97, 101, 133, 168

Terpenoids, 133
Tetrahydrofuran (THF), 208
Texture, 11, 12, 25
  profile analysis, 11, 12
Therapeutic properties, 44
Thermal
  analysis, 12
  behavior, 97
  gelling, 98
  processing, 24
  stability, 103
  treatments, 26
Thermolabile substances, 76
Thiobarbituric acid value, 10, 25
Three-dimensional network, 94, 97
Thymol, 27, 133, 134, 140
Tilapia, 19, 23
Tissue, 19, 66, 76, 94, 95, 103, 185, 202, 214
  pigmentation, 66
Titratable
  acidity, 8
  activity, 12
Torchbearers, 222, 230, 234
Torularhodin, 74
Torularodine, 72
Torulopsis, 22
Toxic
  aldehydes, 22
  metabolites, 48
  metals, 231
Toxicity, 5, 163, 233, 234
Trans-cinnamaldehyde, 18, 27
Transglutaminase, 18
Transition, 5, 69
Tryptophan, 191
Tulsi extract, 12
Tumorigenesis, 67

U

Ulcerative colitis (UC), 48, 49
Umbilical cord, 231, 232
Upregulation, 48
United States Department of Agriculture (USDA), 26

V

Vainilla oleoresin, 13
Vapor transmission rate, 12
Vascular function, 214
Vegetable proteins, 9
Vegetarianism, 9
Vermicompost, 108
Vermicomposting process, 110
Vinyl acetate, 96
Violacein, 211
Violaxantina, 70
Viroids, 191
Virus, 191, 201
Visionary
  coin, 226, 228–230, 232
  paradigm, 225, 232, 239
  scientific arena, 239
  timeframe, 225
Vitamin, 2, 53, 64, 168
  precursors, 66
  synthesis, 47
Volatile, 21, 92, 101, 111, 132, 162, 168, 231
  nitrogenous bases, 21

W

Water
  activity, 25
  content, 25
  holding capacity, 10, 19, 25, 27
  purification, 237, 241
  solubility, 10
  vapor
    permeability (WVP), 10–12, 15, 16, 19, 25, 137, 138, 140
  transmission, 10, 12
Waxes, 2, 5, 6, 9, 13–15, 51
Weight
  loss, 11, 12
  ratio, 100, 213
Wheat, 7, 15, 51, 95, 190, 193
Whey protein, 7, 10, 15, 18, 24, 25, 55, 103, 138

X

Xanthan, 27
Xanthophyllomyces, 65, 73
Xanthophylls, 64, 65, 67, 69
Xylanase, 110, 111

Y

Yersinia enterocolitica, 22

Z

Zea mays, 70
Zeaxanthin, 66–70
Zein (corn), 7, 9, 15, 138
Zygotic
 embryos, 191
 plantlets, 189
 polyembryony, 186, 191
 rootstock, 188